Glossary

of Aquatic Habitat Inventory Terminology

Support for this publication
was provided by

Sport Fish Restoration Act Funds

administered by the

**U.S. Fish and Wildlife Service
Division of Federal Aid**

Glossary

of Aquatic Habitat Inventory Terminology

Neil B. Armantrout

Compiler

Western Division
American Fisheries Society
5410 Grosvenor Lane, Suite 110
Bethesda, Maryland 20814-2199

Suggested Citation Format:

Armantrout, N. B., compiler. 1998. Glossary of aquatic habitat inventory terminology. American Fisheries Society, Bethesda, Maryland.

Library of Congress Catalog Number: 98-87146
ISBN: 1-888569-11-5

Printed in the United States of America on acid-free paper.

Illustration credits:
Cover illustration from Firehock and Doherty 1995.
Illustrations from Helm 1985 drawn by Jennifer Nielsen.

American Fisheries Society
5410 Grosvenor Lane, Suite 110
Bethesda, Maryland 20814-2199, USA

This glossary is dedicated

to the memory of

Dr. William T. Helm.

Contents

Preface ix

Acknowledgments xi

Symbols and Abbreviations xiii

Terms and Definitions 1

References 127

Preface

An aquatic system is dependent upon the hydrological cycle that provides the water supply for the system through the amount, timing, and pattern of precipitation. Water from precipitation moves as surface or groundwater by gravity until it eventually reaches the ocean or enters the atmosphere as water vapor through evaporation or transpiration by plants. Condensation of water vapor at high altitudes results in precipitation and the hydrology cycle repeats itself.

The pattern of water movement on or under the surface of the earth is largely determined by the geomorphology where the porosity and permeability of substrates determine the flow of surface water as well as the infiltration and flow rate of groundwater. Surface patterns of standing and flowing water develop in response to landscape features that, in turn, help to form those features through erosion and deposition. Water, as surface or groundwater, governs the type and extent of vegetative cover in an area that results in a variety of aquatic and terrestrial habitats. Vegetation modifies the movement and erosive power of water that alters landscape features.

The interaction of climate, geomorphology, and vegetation produces a variety of aquatic features, ranging from rivers and lakes to wetlands and springs. The surface features do not exist as separate entities, but represent the most visible features of the much more extensive integrated aquatic system. Our attention is given most often to the largest, most visible features of an aquatic system although they may represent only a small fraction of the overall system.

Recently, humans have played an increasing role in the distribution and uses of water that have altered most major aquatic ecosystems on earth. Changes to both surface and subsurface aquatic systems have altered the natural hydrologic processes, landscape features, and even climate. Often humans fail to consider the interactions of the entire hydrologic cycle, climate, geomorphology, and vegetation, as well as the impacts on various habitats and organisms that use such habitats.

Background and Coverage

Existing aquatic and terrestrial systems are the result of dynamic interactions between water, climate, geology, vegetation, and human influences with time. Aquatic systems occur in a variety of types that range from flowing systems (e.g., rivers and streams) to standing water (e.g., wetlands, marshes, ponds, lakes, and reservoirs). Although influenced by various factors, aquatic systems exhibit consistent patterns so that they can be described and classified.

With growing demands for the multiple-use of aquatic resources, the importance of aquatic resource inventories and knowledge of the complex interactions of hydrology, hydraulics, and geomorphology is extremely important to understanding the overall effect of biological, chemical, and physical factors on an entire ecosystem.

While hydrologic processes and aquatic habitats exhibit consistent patterns, terminology and methods for aquatic habitat inventories vary among organizations, agencies, and disciplines. The resulting confusion and misunderstanding limit exchange and comparison of data. Researchers and managers are hindered from sharing data because terminology and methods are not standardized. With the rapid development of computer programs, the need for standardization of inventory systems has grown in concert with increased opportunities for data analysis at the local and landscape scale.

Realizing the need for standardization, the American Fisheries Society developed this glossary of aquatic habitat terminology.

Under the chair of Dr. William Helm, the Western Division of the American Fisheries Society prepared the first draft of a "Stream Inventory Glossary" which was based on terminology from the Western United States and Canada. It was then expanded to cover flowing, standing, and groundwater terminology used throughout North America. Some terms and definitions that are used in other countries and are frequently found in the literature were added for completeness.

This glossary has been developed to encourage consistent and standard use of the terms used by workers who conduct inventories and analyses of aquatic habitats. Standardization of terms and definitions provides a common language for habitat work and ensures the applicability and accuracy of inventory methods.

This document was prepared with a broad perspective on aquatic habitat terminology. Because of the interactive and integrative nature of aquatic systems with the landscape, many terms from the disciplines of meteorology, hydrology, hydraulics, and geomorphology have been included. Inventory and analysis of aquatic ecosystems have evolved from a localized approach to a watershed or landscape perspective and so are intertwined with ecology, meteorology, hydrology, hydraulics, and geology.

Some terms used by different professions and in different geographic regions have developed more than one definition. Where there are differences in meaning between the biological and physical science definitions, both definitions are provided for clarification. A conscious decision was made not to include all terms and definitions that are applicable to vegetation, ecology, physiology, soils, and fish sampling, but to include terms that are often used in inventories of aquatic habitats.

This glossary is part of a larger effort of the American Fisheries Society to produce a manual of standard methods for conducting aquatic habitat inventories and assessments. For this reason, many terms in this glossary do not provide details on how the information related to these terms is collected or used since those details will be described in detail in the techniques manual.

Balon's (1982) concerns about committees and nomenclature were carefully considered in the compilation of this glossary. Balon questioned the use of committees in standardizing nomenclature when he stated: "How much useful contribution, for example, can be expected from committees on 'standardization of nomenclature'? Nearly in all cases I know about, conclusions reached in such committees ratify parochial dogmas, many disproved [a] long time ago, rather than contribute to knowledge. Consequently, by false restriction of choices they retard future contributions." He concluded by stating, "It is not the nomenclature that matters but the clear definitions of the contents given to terms, a truism most frequently misunderstood."

Although this glossary is intended to standardize definitions of terms that are applicable to aquatic systems, no restrictions apply to the evolution and development of new terms that better describe such systems. American Fisheries Society members also intend to apply adaptive management to this glossary where, periodically, the terms in the glossary will be reviewed and refined, or updated, with the best information available at that time.

Acknowledgments

Funding for preparation and publication of this glossary was provided by the U.S. Fish and Wildlife Service, Division of Federal Aid. This compilation began as a project of the Western Division, American Fisheries Society, under the direction of Dr. William Helm. An American Fisheries Society committee with Neil Armantrout as chair continued the effort in 1981 and expanded the glossary to include terms used in inventories of all types of aquatic habitats. Neil Armantrout has been the principal compiler of the present glossary. He has received input from numerous people and has integrated glossary terms and definitions used by other disciplines such as the Geological Society and the Society of American Foresters. Various fishery biologists, including American Fisheries Society members either as individuals or as members of various agencies or organizations from the United States and Canada have graciously donated their time to provide terms and definitions or to review various sections of this glossary and provide suggestions for improvement.

The present glossary of terms and definitions used in conducting inventories of aquatic habitats is the cumulative effort of all persons who provided editorial comments for clarity and consistency and who helped to integrate terms and definitions from various sources listed in the reference section. Appreciation is extended to Mike Aceitano, John Anderson, Richard O. Anderson, Carl Armour, John Bartholow, Ken Bates, David H. Bennett, Joseph Bergen, Eric Bergersen, Ray Biette, N. Allen Binns, Peter Bisson, Michael Bozek, Richard E. Bruesewitz, C. Fred Bryan, Mason Bryant, John W. Burris, Dieter N. Busch, Jean Caldwell, Tom Chamberlin, James Chambers, Chris Clancy, Don Cloutman, Bruce Crawford, David Cross, Peter Delaney, Don Duff, Mark P. Ebener, Russell H. England, David Etnier, Otto Fajen, C. Michael Falter, David Fuller, Barb Garcia, Ronald Garvaelli, Leonard J. Gerardi, Gareth A. Goodchild, Robert S. Gregory, Bob Griffith, Steve Gutreuter, James Hall, Jim Henriksen, George Holton, Wayne Hubert, Mark Hudy, Philip J. Hulbert, Leroy J. Husak, Jack Imhof, Don Jackson, Alan Johnson, Gary B. Kappesser, John Kauffman, Jeff Kershner, John F. Kocik, Tom Lambert, Bill Layher, Steve Leonard, Gerald E. Lewis, Ron Lewis, John Lyons, Mike Maceina, O. Eugene Maughan, Daniel B. McKinley, Michael R. Meador, Keiteh Meals, Larry Mohn, Samuel C. Mozley, Donald Orth, Lewis L. Osborne, Vaugh L. Paragamian, Ron Ptolemy, Mike Purser, Charles F. Rabeni, Ron Remmick, John Rinne, Robert M. Ross, Gordon W. Russell, Robert N. Schmal, Monte E. Seehorn, Paul Seelbach, Mark Shaw, Brian Shuter, E. G. Silas, Dennis Smith, G. D. Taylor, William Thorn, Dennis Tol, Bill Turner, Harold H. Tyus, Richard A. Valdez, Bruce R. Ward, J. C. Wightman, and Richard S. Wydoski. In addition to these individuals, a number of anonymous contributors and reviewers assisted in the compilation and editing of this glossary.

Final editing, design, and production was handled by Beth Staehle, Robert Rand, and Janet Harry in the American Fisheries Society Editorial Office.

Symbols and Abbreviations

The following symbols and abbreviations may be found in this book without definition.
Also undefined are standard mathematical and statistical symbols given in most dictionaries.

°C	degrees Celsius
cm	centimeter
d	day
dL	deciliter
E	east
e	base of natural logarithm (2.71828…)
e.g.	(exempli gratia) for example
et al.	(et alii) and others
etc.	et cetera
°F	degrees Fahrenheit
ft	foot (30.5 cm)
ft^3/s	cubic feet per second (0.0283 m^3/s)
g	gram
gal	gallon (3.79 L)
h	hour
ha	hectare (2.47 acres)
in	inch (2.54 cm)
i.e.	(id est) that is
k	kilo (10^3, as a prefix)
kg	kilogram
km	kilometer
L	liter (0.264 gal, 1.06 qt)
lb	pound (0.454 kg, 454g)
lm	lumen

log	logarithm
m	meter (as a suffix or by itself); milli (10^{-3}, as a prefix)
mi	mile (1.61 km)
min	minute
N	normal (for chemistry); north (for geography); newton
N	sample size
oz	ounce (28.4 g)
P	probability
pH	negative log of hydrogen ion activity
ppm	parts per million
qt	quart (0.946 L)
S	siemens (for electrical conductance); south (for geography)
s	second
U.S.	United States (adjective)
USA	United States of America (noun)
W	watt (for power); west (for geography)
yd	yard (0.914 m, 91.4 cm)
μ	micro (10^{-6}, as a prefix)
°	degree (temperature as a prefix, angular as a suffix)
%	per cent (per hundred)
‰	per mille (per thousand)

Terms and Definitions

abandoned meander channel   See *abandoned meander channel* under **channel pattern**.

abiogenic   Not derived from living organisms.

abioseston   Nonliving components of the seston. See *seston* and **tripton.**

abiotic   Nonliving components in the environment.

ablation   (1) Wearing away of the surface of rocks or of glaciers by the kinetic energy of running or dropping water. (2) All the processes, such as melting or evaporation, by which ice or snow becomes transformed into water or water vapor.

ablation moraine   See *ablation moraine* under *moraine.*

ablation till   The coarse material that accumulates on top of a melting glacier.

ablation zone   Lower part of a glacier where annual water loss exceeds snow accumulation.

absorption   The selective uptake of fluids or substances that are in solution. Absorption may occur in organic matter or by cells in living organisms.

abstraction   (1) Permanent removal of surface flow from a stream channel. (2) Shifting of a stream from one basin to another.

abutment   The two support ends of a bridge that connect it to the adjoining land mass or road fill.

abyssal depth   (1) Maximum depth of a lake or sea. (2) Great depth in oceans or lakes where light does not penetrate. Generally applied to the marine environment.

abyssal zone   The bottom stratum of water. Generally applied to the marine environment.

accelerated erosion   See *accelerated erosion* under *erosion.*

acclimatize   (1) To adapt to a new temperature, altitude, climate, environment, or situation. (2) To become habituated to an environment where a plant or animal species is not native.

accretion   (1) Gradual accumulation of flow from seepage of water. (2) The accumulation of sand, silt, and gravel in a stream channel. (3) Process, driven by tectonics, whereby the continental margin grows by addition of ocean crust and sediment at a subduction zone.

accumulation zone   Part of a glacier where snow accumulation exceeds melting and evaporation.

acid drainage   Runoff water with a pH less than 7.0, with total acidity exceeding total alkalinity, often associated with discharge from mining or reclamation operations.

acidic deposition   The process by which acids are deposited into the environment through rain, sleet, or snow, when the pH of the precipitation is less than 7.0.

acidification   An increase in the acidity of an ecosystem, caused by natural processes or acid deposition.

acidity   (1) Quantitative capacity to neutralize a base to a designated pH. (2) A measure of hydrogen ion concentration, with a pH less than 7.0. See *pH.*

acid rain   In general, rain having a pH of less than 7.0 can be considered acidic. Acid rain, as used in the literature, has a pH that is low enough (e.g., 5.6) to potentially threaten the life or well-being of vegetation or aquatic systems. See *pH.*

acquic moisture regime   A seasonal moisture condition in soil that is virtually free of dissolved oxygen because the soil is saturated by groundwater or by water of the capillary fringe.

acre-foot   See *acre-foot* under **dimensions**.

active channel   (1) Short-term geomorphic feature, defined by the bank break, that marks a change to permanent vegetation. (2) The portion of a channel in which flows occur frequently enough to keep vegetation from becoming established. An active channel is formed and maintained by normal water and sediment processes.

active erosion   See *erosion*.

active slope   Slope of a hill or mountain, typically exceeding 45%, where detrital materials accumulate behind obstructions from valley incision processes.

active valley wall processes   All currently active processes causing the movement of materials on valley walls or relic terraces, even if they do not reach the stream channel.

   bench processes   Processes that extend onto a bench or terrace but do not reach the active floodplain or valley flat.

   floodplain processes   Processes that extend out onto the floodplain or flat valley.

   mountain processes   Processes restricted to mountainsides without a connection to the valley bottom.

active water table   Condition in which the soil saturation zone fluctuates and results in periodic anaerobic soil conditions.

actual shoreline   Line formed by the water interface wherever it covers the bed of a body of water at the mean high water level.

acute meander   See *acute meander* under **meander**.

adaptation   Ability of an organism to adjust to changes in environmental conditions including biological, chemical, and physical factors.

adfluvial   (1) Life history strategy in which adult fish spawn and juveniles subsequently rear in streams but migrate to lakes for feeding as subadults and adults. (2) Pertaining to flowing water. (3) Movement toward a river or stream. See *fluvial*.

adjacent   See *adjacent* under **wetlands**.

adsorbed water   Water that condenses or accumulates on a surface.

adsorption   Adhesion of an extremely thin layer of gas molecules, dissolved substances, or liquids to a solid surface that they contact.

aeolian lake   See *aeolian lake* under **lake**.

aerated pond or lagoon   See *aerated pond or lagoon* under **pond**.

aeration   To enhance the oxygen concentration of water.

aerial photography   See *aerial photography* under **remote sensing**.

aerobic   (1) Zone in a body of water where free oxygen is present. (2) Life processes occurring only in the presence of molecular oxygen.

aestival lake   See *aestival lake* under **lake.**

aestival pond   See *aestival pond* under **pond**.

affluent   Water flowing to a standing water body.

aggradation   Geologic process in which inorganic materials carried downstream are deposited in streambeds, floodplains, and other water bodies resulting in a rise in elevation in the bottom of the water body. Compare with *degradation*.

aggraded channel   See *aggraded channel* under **channel geometry**.

aggregate   Granular mineral materials such as sand and gravel. See also *aggregate* under **large organic debris**.

aggregate stability   Erodibility of soil in terms of its tendency to separate when wetted.

a-jacks   See *a-jacks* under **habitat enhancements**.

albedo   Reflecting power of a surface, expressed as the reflected fraction of the radiation incident at the surface.

alcove   See *alcove* under *slow water, pool, scour pool* under the main heading **channel unit**.

algal bloom   Proliferation of one or several species of phytoplankton to high cell densities during favorable environmental conditions.

algal wash   Shoreline drift composed mainly of filamentous algae.

alkali   A substance which neutralizes an acid, usually a base such as metallic hydroxide.

alkali lake   See *alkali lake* under **lake**.

alkaline   In general, any water or soil with a pH greater than 7.0. More commonly used for a pH greater than 7.4 that promotes accumulation of salts and affects the physiology of plants and animals. See *pH.*

alkalinity   Acid-neutralizing capacity of water usually due to carbonates, bicarbonates, and hydroxides; expressed as mg/L of $CaCO_3$.

allochthonous   Exogenous food organisms, organic matter, and nutrients originating outside and transported into an aquatic system.

allogenic   Refers to origination that is external to a system.

alluvial   Related to material deposited by running water.

alluvial cone   Material carried by ephemeral streams from higher elevations and accumulating at the mouth of a gorge as a moderately steep conical formation that slopes equally in all directions.

alluvial dam   A dam formed by the accumulation of sediment that blocks the channel of a stream.

alluvial deposit   Clay, silt, sand, gravel, or other sediment carried by flowing waters and deposited when the water velocity drops below that required to keep the material in suspension or move the bed load. Synonymous with *alluvial fill*.

alluvial fan   Fan-shaped area of deposits of sediments formed by a stream where it exits a valley onto a floodplain and results from an abrupt decrease in water velocity.

alluvial fill   Alluvial material that accumulates as a result of a reduction in water velocity, thereby allowing the alluvial material to settle. Synonymous with *alluvial deposit*.

alluvial flat   (1) An alluvial valley and the relatively low-gradient stream reaches associated with it. (2) Term used to refer to an isolated alluvial valley that is bound upstream and downstream by steeper-gradient canyon stream segments.

alluvial plain   An expanse of land formed, at least in part, by deposited materials through which an alluvial stream meanders. Usually applied to the gentle slope in a coastal plain.

alluvial soil   A soil consisting of recently deposited sediment and showing no horizon development or other modification of the deposited material.

alluvial stream   Named after the silts, clays, sands, and gravels of river origin that compose the bed, banks, and floodplains of streams. Alluvial streams tend to be large, composed of bed materials conveyed from upstream, and are characterized by a distinctive S-shaped channel pattern that is free to shift slowly (i.e., meander) in the valley. See also *alluvial stream* under *stream*.

alluvial terrace   A relatively level plain or step that is formed by the deposition of alluvial sediments in floodplains when water velocities are reduced as the river subsides from a high flow event.

alluvial valley   A valley form which resulted from deposition of relatively large stores of alluvium by fluvial erosion processes.

alluvial valley floor   Unconsolidated stream-laid deposits forming the valley floor.

alluvion   A gradual increase of land on a shore or streambank by the action of water, from natural or artificial causes.

alluvium   A general term for all deposits resulting, directly or indirectly, from the sediment transport of streams that is deposited in streambeds, floodplains, lakes, and estuaries. See also *alluvium* under *valley segments*.

alpha-mesosaprobic zone   See *alpha-mesosaprobic zone* under *saprobien system*.

alternate bar   See *alternate bar* under *bar*.

ambient   General conditions in the environment.

amictic   See *amictic* under *mixing*.

amplitude   See *amplitude* under *meander*.

anadromous   A life history strategy of fishes, that includes migration between fresh- and saltwater, in which reproduction and egg deposition occurs in freshwater while rearing to the adult stage occurs in the ocean. Compare with *catadromous, diadromous, oceanadromous, potamodromous*.

anaerobe   An organism that lives where free oxygen is absent.

anaerobic   (1) Environmental conditions where free oxygen is absent. (2) Life processes that occur in the absence of molecular oxygen.

analytical watershed   Term applied to a drainage basin for analyzing cumulative impacts on natural resources.

anastomizing channel   See *anastomizing channel* under *channel pattern*.

anchialine pool   See *anchialine pool* under *pond*.

anchor ice   Ice formed on substrate or objects beneath fresh- or saltwater surfaces when the water becomes supercooled.

angle of repose   Maximum slope or angle at which a material will remain stable without sliding or slumping.

annual maximum daily discharge   See *annual maximum daily discharge* under *discharge*.

annual maximum instanteous discharge   See *annual maximum instanteous discharge* under *discharge*.

annual mean discharge   See *annual mean discharge* under *discharge*.

annual minimum daily discharge   See *annual minimum daily discharge* under *discharge*.

anoxic   Lack of oxygen.

antagonism   Interaction of two or more substances (e.g., chemicals) such that the action of any one of them on living cells or tissues is lessened.

antecedent soil moisture   Degree of wetness of soil at the beginning of a runoff or storm period, expressed as an index or as the total volume of water stored in the soil.

anthropogenic   See *anthropogenic* under *streambank material*.

antidune   See *sand wave*.

antinode   See *antinode* under *wave*.

aphotic   Depth of a body of water where light does not sufficiently penetrate to allow photosynthesis.

aphotic zone   Stratum of a water body below which sufficient light is not present to allow photosynthesis.

aphytal   (1) Refers to an area that lacks plants. (2) The plantless (profundal) zone of a lake bottom.

apparently stable channel   See *apparently stable channel* under *channel pattern*.

apparent velocity   See *interstitial velocity*.

apron   (1) A protective structure in a stream used to prevent erosion around an artificial structure. (2) A pavement constructed at the lower side of a dam wall to protect the foundation from erosion.

aquamarsh   See *aquamarsh* under *wetlands*.

aquatic   Term applied to growing, living in, frequenting, or pertaining to water.

aquatic bed   Wetlands and deeper water habitats dominated by plants that grow principally on or below the surface of the water for most of the growing season in most years. See also *aquatic bed* under *wetlands*.

aquatic ecosystem   Any body of water, such as a wetland, stream, lake, reservoir, or estuary that includes all organisms and nonliving components, functioning as a natural system. See *aquatic system*.

aquatic habitat   A specific type of area with environmental (i.e., biological, chemical, or physical) characteristics needed and used by an aquatic organism, population, or community.

aquatic system   Any body of water, including lakes, reservoirs, streams, springs, or estuaries with associated living organisms and all nonliving components that function as a natural system. See *aquatic ecosystem*.

aquic   Soils that have continuous or periodic saturation and reduction. These conditions are indicated by redoximorphic features.

aquiclude   A poorly permeable underground bed, formation, or group of formations, often saturated, that impedes groundwater movement, and does not yield water freely to a well or spring. However, an aquiclude may transmit appreciable water to or from adjacent aquifers, and where sufficiently thick, may constitute an important groundwater storage unit.

aquic moisture regime   A mostly reducing seasonal soil moisture regime that is free of dissolved oxygen due to saturation by groundwater or its capillary fringe.

aquifer   An underground bed or layer of sand, earth, gravel, or porous stone that contains water or permits its passage.

arched roots Roots produced on plant stems in a position above the normal position of roots that serve to brace the plant during and following periods of prolonged inundation in water.

archival reach See *archival reach* under **reach**.

arc of the sun See *arc of the sun* under **solar radiation**.

area See *area* under **dimensions**.

area–capacity curve A graph showing the relation between the surface area of the water in a reservoir, the corresponding volume, and elevation.

argillorheophilic See *argillorheophilic* under **benthos**.

arm A long and relatively narrow water body extending inland from the main body of a lake or reservoir, and usually associated with a major tributary. Generally the term "arm" is applied to a water area that is greater in length and narrower in width than a bay or a cove. See **bay, cove, embayment**.

armoring (also armouring) (1) Formation of an erosion-resistant layer of relatively large particles on the surface of a streambed or streambank, resulting from removal of finer particles by erosion, which resists degradation by water currents. (2) Application of materials to reduce erosion.

arroyo A water-carved channel or gully in an arid area, usually rather small with steep banks and flat bottom, dry much of the time due to infrequent rainfall and the shallowness of the cut which does not penetrate below the level of permanent groundwater. In Mexico, the term "arroyo" often refers to permanent streams as well.

artesian well A well which continues to flow without a pumping mechanism as a result of groundwater pressure.

articulated concrete mattress See *articulated concrete mattress* under **habitat enhancements**.

articulation (1) Connection between two or more bodies of water. (2) Ratio of total area of inlets and bays to the total area of a lake or reservoir.

artificial beach Shore of a water body where the substrate or physical contours have been altered by human design or activity for recreational or aesthetic purposes.

artificial channels See *artificial channels* under **habitat enhancements**.

artificial control A weir or other constructed structure which serves as the control for stream-gaging stations, fish passage facilities, streambed retention, or protection against erosion.

artificial holes See *artificial holes* under **habitat enhancements**.

artificial meander See *artificial meander* under **habitat enhancements**.

artificial recharge Recharge of groundwater aquifers by pumping or seepage of water introduced by human activity.

artificial reef See *artificial reef* under **habitat enhancements**.

artificial riffle See *artificial riffle* under **habitat enhancements**.

artificial substrate A device placed in the water that provides living space for aquatic organisms; most often used for biomonitoring.

artificial wetland See *artificial wetland* under **wetlands**.

aspect Direction a slope faces with respect to the cardinal compass points.

assimilation Ability of a water body to absorb materials and substances to purify itself.

assimilation capacity (1) Capacity of a natural water body to receive wastewaters, without deleterious effects, or toxic materials, without damage to aquatic life or humans that consume the water. (2) To incorporate and convert waste waters without deleterious effects. (3) Biological Oxygen Demand (BOD), within prescribed dissolved oxygen limits.

association (1) Plant and animal communities of a particular kind that are consistently found together. (2) A group of plant and animal species.

astatic Water bodies in which surface levels fluctuate.

   perennial astatic Water body that rises and falls but does not dry up every year.

   seasonal astatic Water body that dries up annually.

attenuation (1) Reduction in light intensity in water because of absorption and scattering by water molecules, suspended particles, and dissolved substances. (2) Reduction of the peak of a hydrograph by natural or artificial water storage.

attribute See *habitat component* and *attribute* under *remote sensing*.

aufwuchs Attached microscopic organisms growing on the bottom or other substrates.

autochthonous Endogenous materials such as nutrients or organisms fixed or generated within the aquatic system.

autotrophic Organisms that manufacture their own food using carbon dioxide or other inorganic substances. See also *autotrophic* under *trophic*.

available water capacity Ability of a soil to hold water in a form available to plants that is generally expressed in inches of water per inch of soil depth.

avalanche See *avalanche* under *landslide*.

avalanche cone See *avalanche cone* under *landslide*.

avalanche track See *avalanche track* under *landslide*.

average annual discharge See *average annual discharge* under *flow*.

average annual inflow See *average annual inflow* under *flow*.

average depth See *average depth* under *dimensions*.

average linear velocity See *average linear velocity* under *groundwater*.

average width See *average width* under *dimensions*.

AVHRR See *AVHRR* under *remote sensing*.

avulsed lands Lands that have been uncovered by a relatively sudden change in alignment of a river channel or by a comparable change in some other water body. Such lands remain as uplands or islands after the change.

avulsion An abandoned stream channel resulting from movement of the channel to a new route or direction.

azimuth An angle measured clockwise from a meridian, going north to east. See also *azimuth* under *remote sensing*.

▶ b

backbar channel See *backbar channel* under *channel pattern*.

backflow Reversal in the direction of water flow through a conduit or channel in the direction opposite to normal flow.

backflow billabong See *backflow billabong* under *billabong*.

backshore Zone of a typical beach profile above mean high water. Also used for the zone covered with water only during exceptionally severe storms.

backslope Steep slopes formed by bank erosion from flowing water.

backswamp See *backswamp* under *wetlands*.

backwash (1) Water thrown backward by the motion of a boat or other water craft. (2) Flow of water returning to a water body that occurs after the passing of a wave.

backwater (1) Water backed up or retarded in its course as compared with its normal or natural condition or flow. (2) A naturally or artificially formed arm or area of standing or slow-moving water partially isolated from the flow of the main channel of a river. (3) An area off the main water body of lakes, reservoirs, and bayous. (4) In stream gaging, a rise in stage produced by a temporary obstruction such as ice or weeds, or by flooding of the stream below. The difference between the observed stage and that indicated by the stage-discharge relation is reported as backwater. (5) Seasonal or permanent water bodies found in the lowest parts of floodplains, typically circular or oval in shape. (6) The plural (*backwaters*) is often used to refer to the upper, more riverine portion of a reservoir. See also *backwater* under *slow water*, *pool*, *dammed pool* with the main heading of **channel unit**.

backwater curve (1) Term is used in a general sense to denote calculated or measured water surface profiles, or for the profiles where the water depth has been increased by placement of an obstruction such as a dam or a weir. (2) A water surface profile in a section of stream in which there is a gradual transition in the depth of flow upwards or downwards relative to the channel bottom. (3) Change in elevation in a reservoir with increasing distance from the dam.

backwater deposits Overbank deposits of fine sediments deposited in slack water of a flood-plain or outside natural levees.

backwater effect The effect of a dam or other obstruction in raising the upstream water surface.

backwater flooding Overbank inundation (i.e., flooding) that occurs during high runoff from a nearby stream.

badlands Generally barren, rough, and broken land that is strongly dissected or gullied by a fine drainage network with a very high drainage density. Most common in semi-arid regions where streams have entrenched into soft erodible materials.

baffle See *baffle* under ***habitat enhancements***.

bajada A broad, gently inclined, piedmont slope formed by lateral coalescence of a series of alluvial fans with a broadly undulating trans-verse profile, resulting from the shape of the fans.

ballena A landform of round-topped ridgeline remnants of fan alluvium. Generally, the broadly rounded ridge shoulders meet to form a narrow crest and merge smoothly with the concave backslopes.

band See *band* under ***remote sensing***.

band ratio See *band ratio* under ***remote sensing***.

bank See ***streambank***.

bank angle Angle formed by a downward sloping bank, measured in degrees from the horizontal such that a vertical bank is 90°. If the angle is variable, it is calculated by using the average of three measurements.

bank depth See *bank depth* under ***dimensions***.

bank erosion Erosion of bank material caused by water current, wave action, or surface erosion.

bank-full depth Depth of water measured from the surface to the channel bottom when the water surface is even with the top of the streambank.

bank-full discharge Maximum streamflow that can be accommodated within the channel with-out overtopping the banks and spreading onto the floodplain. Generally the level associated with two- or three-year streamflow events.

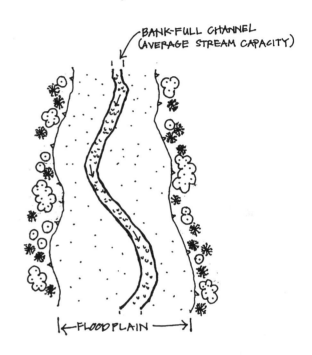

bank-full stage (from Firehock and Doherty 1995)

bank-full stage Stream stage where stream reaches bank-full depth.

bank-full width Channel width between the tops of the most pronounced banks on either side of a stream reach.

bank height Distance between the channel bed and the top of the bank.

bank revetment See *bank revetment* under ***habitat enhancements***.

bank sloughing Slumping of saturated, cohesive soils surrounding a water body that are incapable of free drainage during rapid drop in water level.

bank stability Pertains to the resistance of a bank to erosion.

   failing bank A bank that is unable to maintain its structure because of active erosion.

stable bank A bank that, even with a steep slope, has no evidence of active erosion, breakdown, tension cracking, or shearing.

unstable bank A bank that shows active erosion or slumping.

bank stabilization See *bank stabilization* under ***habitat enhancements***.

bank storage Water absorbed and stored in the voids in the soil cover in the bed and banks of a stream, lake, or reservoir, and returned in whole or in part as the surface drops.

bank width See *bank width* under ***dimensions***.

bar A submerged or exposed ridge-like accumulation of sand, gravel, or other alluvial material formed in a lake or in the channel, along the banks, or at the mouth of a stream where a decrease in velocity induces deposition.

  alternate bar Bars (i.e., depositions) that change from one side to the other of a winding thalweg.

  braided bar Pattern of river bars with numerous interconnected small channels at lower flows that form in streams with high volumes of bed material.

  channel junction bar Bar formed where two channels intersect.

  cross-over bar See *transverse bar* under ***bar***.

  delta bar A bar formed immediately downstream of the confluence of a tributary and the main stream. Compare with *junction bar* under ***bar***.

  diagonal bar A bar that forms diagonally to a stream channel. Compare with *tansverse bar* under ***bar***.

  diamond bar A form of braiding in which the multiple interconnected channels form mid-channel bars.

  diamond-braided bar Multiple diamond-shaped interconnected mid-channel bars that are characteristic of braided rivers.

  dune bar Wave-like streambed formation that commonly occurs in relatively active channels of sand streambeds.

  islands Exposed bars or land segments within the stream channel that are relatively stable and normally surrounded by water.

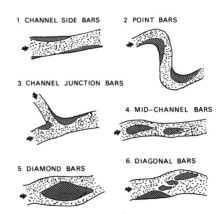

types of bars (adapted from Kellerhalls et al. 1995 with permission of ASCE)

  junction bar A bar formed at the junction of two streams, usually because sediment transported by a tributary is deposited in the slower-moving water of the mainstream. Compare with *delta bar* under ***bar***.

  lee bar A bar caused by eddies and lower current velocities and formed in the lee of large immovable objects such as boulders or logs.

  mid-channel bar Bar formed in the mid-channel zone, not extending completely across the channel. Also called a middle bar or mid-bar.

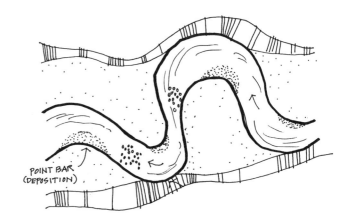

point bar (from Firehock and Doherty 1995)

  point bar Bar found on the inside of meander bends.

  reattachment bar Bar that extends from the downstream end of a recirculating eddy in large rivers and separated from the streambank by a recurrent channel. Compare with *separation bar* under ***bar***.

  separation bar Bar formed that extends from the upstream end of a recirculating eddy of large rivers. Compare with *reattachment bar* under *bar.*

side bar   Bar located at the side of a river channel, usually associated with the inside of slight curves. Also called a lateral bar.

transverse bar   Bar that extends diagonally across the full width of the active stream channel. Compare with *diagonal bar* under *bar.*

bar and channel   A pattern found in younger floodplains and terraces, consisting of ridgelike bars of accumulated coarse sediments and channels of finer textures. Maintained by the competence of the stream, the features become less pronounced as higher lying bars erode.

bar presence   Refers to the relative abundance of bars or islands in a river channel.

barrage   Dams or weirs used for creating barriers to fish movements and facilitate capture of fish. Also used as a general term for dams or weirs.

barren shoal   Shallow lake bottom with few or entirely devoid of aquatic plants.

barrier   Any physical, physiographic, chemical or biological obstacle to migration or dispersal of aquatic organisms.

barrier beach   An exposed ridge of deposits separated from the mainland by water.

barrier dam   Low dam across a stream to divert water or to guide fish to the entrance of a fish-way. Barrier dams may be used to block up-stream migrations of fish.

barrier flat   A relatively flat area, often with pools of water, separating the exposed or seaward edge of a barrier from the lagoon behind it.

barrier island   A long broad sandy island lying parallel to a shore that is built up by the action of waves, currents, and winds and that protects the shore from the effects of the ocean.

bar screen   A device for removing larger materials from effluent.

base flow   See *base flow* under *flow.*

base level   The level to which a stream channel profile has developed and below which signifi-cant erosion by water does not proceed.

baseline   (1) Reference point for comparison of subsequent measurements. (2) Level of a receiv-ing stream.

basin   A topographic area of a watershed or geological land area that slopes toward a com-mon center or depression where all surface and subsurface water drains. See *drainage area.*

basin dam   A dam with a concave top that is built across the mouth of a gorge to prevent formation of an outwash fan.

basin slope   See *basin slope* under *dimensions.*

bathile   Term that pertains to the bed of a lake that is deeper than 25 m.

bathyal zone   Pertaining to deep waters, especially between 183 and 1,830 m (100 and 1,000 fathoms).

bathylimnion   See *bathylimnion* under *stratification.*

bathymetric map   A map depicting the depth contours of the bottom of any water body.

bathymetry   Refers to measurement of depth in a water body.

bay   A water body that forms an indentation in a shoreline, larger than a cove, and typically wider than an arm. Also, a large embayment. See *arm, cove, embayment.*

bay head   That portion of a bay which lies farthest inland from the main water body.

bay mouth   Location where a bay joins a main water body.

bayou   (1) A bay, inlet, backwater, river channel slough, oxbow lake, or channel in coastal marshes and sluggish creeks, or arm, outlet, or tributary of a lake or river. (2) Any stagnant or sluggish creek or marshy lake. Term is most often used in southeastern United States.

beach   The gently sloping zone of demarcation between land and water of lakes or other large water bodies that is covered by sand, gravel, or larger rock fragments.

beach nourishment   Aggradation of a beach from deposition of dredged or fill material placed to replenish eroded areas.

beach ridge   A low, generally continuous area of dune material accumulated by the action of waves and currents that roughly parallel the shoreline and are not influenced by normal tides and storms.

beach terrace   Flat benches, scarps, or terraces formed by waves from well-sorted sand and gravel of lacustrine or marine origin.

beaded stream   A stream consisting of a series of small pools or lakes connected by short stream segments. Commonly found in a region of paternoster lakes, an area underlain by permafrost, or a recently glaciated area. See also *beaded stream* under **stream**.

beaver dam   See *beaver dam* under *slow water, pool, dammed pool* under the main heading **channel unit**.

beaver pond   See *beaver pond* under *pond*.

bed   Bottom of a lake, pond, or other water body.

bedding   Stratification of sedimentary material parallel to the original surface of deposition or inclined to it.

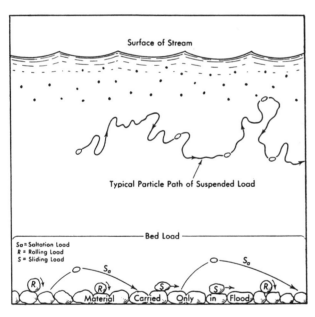

bed load (from Meehan 1991)

bed load   (1) Substrate moving on or near a streambed and frequently in contact with it. (2) See *bed load* under **sediment load**.

bed load discharge   Quantity of bed load passing a given point in a unit of time, expressed as dry weight. See **transport velocity**.

bed load transport rate   A transport rate (Q_b) that is a function of the difference between the effective basal shear stress and the critical shear stress.

$$Q_b = kW(\tau' - \tau_{cr})^{1.5};$$

τ' = basal shear stress;
τ_{cr} = critical shear stress;
W = channel width;
k = a constant.

bed material   Substrate mixture of a stream or lake bed that remains after moderate to high streamflow conditions. In alluvial streambeds, materials that are likely to be moved during moderate to high flows that lead to continued replacement through scour and deposition.

bed material load   See *bed material load* under **sediment load**.

bedrock   Solid rock, either exposed at surface or underlying surficial deposits. See also *bedrock* under **streambank material** and **valley segments**.

bedrock control   The percentage of permanent and stable pools in a river reach that are formed by the presence of bedrock.

bedrock landslide   See *bedrock landslide* under *landslide*.

bedrock type   The parent material of bedrock (e.g., granite or sandstone) in a channel.

bed roughness   A measure of the irregularity of a streambed as it contributes to resistance of flow. Commonly measured in terms of Manning's roughness coefficient. See **Manning's n**.

beheaded stream   See *beheaded stream* under **stream**.

belt of meander   See *belt of meander* under **meander**.

bench   (1) Shelf-like areas in lakes and reservoirs with steeper slopes above and below, developed on horizontal or gently inclined rock layers where overlying softer and less resistant materials have been scoured away. (2) A series of level step-like areas remaining in a floodplain as a result of periodic deposition and erosion. See **terrace**.

benched gully side   See *benched gully side* under **gully side form**.

benchmark   A permanent object of known elevation that is generally located or placed where there is the least likelihood of being disturbed and used as a standard or reference for various physical measurements.

bench processes See *bench processes* under *active valley wall processes*.

bend A curve in the river channel. Term is most often used for an extended curvature of a large river or where the flow of a river changes direction.

beneficial use In water use law, reasonable use of water for a purpose consistent with the laws and best interest of people. Such uses vary by state law and include, but are not limited to: instream, out of stream, groundwater uses, domestic, municipal, industrial water supply, mining, irrigation, livestock watering, fish and aquatic life, wildlife, fishing, water contact recreation, aesthetics and scenic attraction, hydropower, and commercial navigation.

benthic organic matter Undifferentiated particles of organic matter desposited on the bottom of a water body.

> **fine benthic organic matter** Benthic organic matter that is 0.45 μm–1 mm.

> **large benthic organic matter** Benthic organic matter that is greater than 1 mm.

> **nonwoody benthic organic matter** Benthic organic matter from other than woody vegetation, such as sedges or leaf litter.

benthic zone The bottom or bed of a water body.

benthos Bottom-dwelling organisms including plants, invertebrates, and vertebrate animals that inhabit the benthic zone of a water body.

> **argillorheophilic** Benthos inhabiting mostly clay substrates, usually sessile and burrowing organisms.

> **lithorheophilic** Benthos inhabiting solid substrates in flowing water, mainly composed of insects.

> **pelephilic** Benthos inhabiting silty substrates in still waters.

> **pelorheophilic** Benthos inhabiting silty areas in flowing water.

> **phytophilic** Benthos inhabiting backwaters rich in plants.

> **psammorheophilic** Benthos inhabiting sandy bottoms in flowing water, mainly composed of protozoans and small arthropods.

berm See *berm* under *habitat enhancements*. Also compare with *dike*.

beta-mesosaprobic zone See *beta-mesosaprobic zone* under *saprobien system*.

bight A broad, gradual bend or curve in a shoreline that is generally applied to the marine environment.

billabong An Australian term for an isolated pool of water.

> **backflow billabong** A backwater in a floodplain of a river that forms a stagnant pool.

> **channel billabong** A pothole or lagoon in the dry channel of a stream.

biocoenosis The plants and animals that comprise a community.

biocriteria See *biocriteria* under *biological indices*.

biofacies Sedimentary materials of organic origin.

biogenesis (1) Production of living organisms from other living organisms. (2) Materials produced by living organisms necessary for the continuation of life processes. The adjective "biogenic" applies to such production.

biogenic meromixis See *biogenic meromixis* under *mixing*.

biogeographic region Any region that is delineated by its biological and geographic characteristics.

biological accumulation Gradual biological process by which persistent substances accumulate in individual organisms, or through a succession of organisms from primary producers to top carnivores. Also referred to as bioaccumulation.

biological indices A measure of the health or condition of a water body based upon values for specific biological or physical parameters.

> **biocriteria** Biologically based standards used to assess or regulate conditions of a water body.

> **biomonitoring** Use of biological attributes of a water body to assess its environmental health or condition.

biotic condition index (BCI) The community tolerance quotient potential calculated for natural macroinvertebrate, physical, and chemical characteristics divided by the measured community tolerance quotient and multiplied by 100. Values above 90 are excellent, 80–90 are very good, 72–79 are fair, and below 72 are poor.

community tolerance quotient (CTQ) A value that represents a total of the tolerance quotients per sample divided by the number of taxa in the sample. Values generally range from 40 to 108, with the higher number indicating a more tolerant community. Stressed conditions may be shown depending upon the capability and potential of a stream.

diversity index A numerical value derived from the number of individuals per taxon (evenness) and the number of taxa present (richness).

dominance and taxa index (DAT) A diversity index that combines the number of taxa present and the relative dominance of one or more taxa in the samples. Dominance by one or more species (taxa) often indicates a habitat imbalance causing stress on the biological community. The number of species also reflects the condition of the aquatic habitat. A relative value scale for DAT is 18–26, excellent; 11–17, good; 6–10, fair; and 0–5, poor.

index of biotic integrity (IBI) A measure of the degree to which water resource quality deviates from that expected at relatively undisturbed sites. It is calculated based on data from the study of the entire fish community. Its measures, or metrics, fall into three broad categories: species composition, trophic composition, and fish abundance and condition.

tolerance quotient (TQ) A numerical value used to develop an index of the relative tolerance of a taxon to natural environmental levels of physical and chemical parameters found limiting to some species. Low numbers indicate nontolerant species and higher numbers indicate more tolerant species.

biological legacies See *biological legacies* under *large organic debris*.

biological oxygen demand (BOD) (1) The dissolved oxygen required to oxidize inorganic chemicals in water. (2) A measure of oxygen consumption during a fixed period of time. (3) The amount (milligram per liter) of molecular oxygen required to stabilize decomposable organic matter by aerobic biochemical action.

biomass (1) The total weight, at a given time, of living organisms of one or more species per unit area, or of all the species of a community. (2) The weight of a taxon or taxa per unit of surface area or volume of water expressed in units of living or dead weight, dry weight, ash-free weight, or nitrogen content.

biome An extensive complex or community of organisms occurring together. More specifically, a major ecological community such as a desert, grassland, or boreal forest.

biomonitoring See *biomonitoring* under *biological indices*.

biota Refers to all plant and animal life in an area or region.

biotic (1) Pertaining to life or living organisms. Also pertains to being caused by, produced by, or comprising living organisms. (2) Environmental components having their origins in living organisms.

biotic condition index (BCI) See *biotic condition index (BCI)* under *biological indices*.

bioturbation Disturbance of mud, water, or any other medium by the actions of organisms living in or on it.

biozone Zone capable of supporting living organisms.

bitter lake See *bitter lake* under *lake*.

bitters The saturated brine solution remaining after precipitation of sodium chloride during the solar evaporation process.

blanket bog See *blanket bog* under *wetlands*.

blind drain A ditch or trench, partly filled with large stones and covered with earth or brushwood. See also *french drain*.

blind lake See *blind lake* under *lake*.

bloom A period of vigorous plant growth, usually algae, or the aggregation of alga. See *algal bloom*.

blowdown   See *blowdown* under *large organic debris*.

bluff   A cliff, hill, or headland with a broad, steep face. See *cliff*.

boat basin   A protected anchorage for small water craft with facilities for launching and loading.

boathouse   An enclosed structure on land or over water to store (house) water craft.

boat ramp   An artificial sloping structure on the shore of a water body used for boat launching.

bog   See *bog* under *wetlands*.

bole   See *bole* under *large organic debris*.

bolson   An internally drained intermountain basin with nearly flat alluvial plain, playa-like depressions, and individual or coalesced alluvial fans.

bordering land   Land bordering a water body.

borrowpit pond   See *borrowpit pond* under *pond*.

bosque   A small wooded area.

bottom   The ground surface underlying a body of water.

bottomland   A lowland, usually highly fertile, along a stream such as an alluvial floodplain. Also referred to as bottom land.

bottomset bed   Flat-lying bed of fine sediment deposited in front of a delta that is buried by continuous growth of a delta.

bottom slope   The change in the average elevation of a streambed between two cross sections, divided by the distance between them.

boulder   A substrate particle larger than 25 cm (10 in) in diameter. Compare with other substrate sizes under *substrate size*.

boundary layer   Layer where a fluid in contact with a solid surface is no longer influenced by that surface.

brackish   Water with a salt content greater than freshwater but less than seawater.

braided   Refers to a stream that divides into an interlacing network of several branching and reuniting channels separated from each other by islands or channel bars.

braided bar   See *braided bar* under *bar*.

braided channel   See *braided channel* under *channel pattern*.

branch   A term sometimes applied to a small stream or tributary.

breaching   Term applied to a break at the head of a side channel or side slough. May also be applied to a break in a berm, dike, levee, or revetment.

> **controlling breaching**   Breaching condition in which the main stem discharges are equal to or greater than the main stem discharge required to directly govern the hydraulic characteristics within a side slough or side channel. This condition can be denoted as equalling the segment of the flow rating curve, beginning with the point of inflection and beyond.

> **initial breaching**   The main stem discharge that represents the initial point when main stem water begins to enter the upstream end of a side channel or slough.

> **intermediate breaching**   The range of main stem discharges representative of the conditions between the discharges that occur in *initial breaching* and *controlling breaching*.

breadth   See *breadth* and *width* under *dimensions*.

breaker   See *breaker* under *wave*.

breakline   An area characterized by rapid changes in depth, water temperature, water chemistry, water clarity, structure, or cover.

breakwall   See *jetty*.

breakwater   Structural features placed in position to intercept waves and to dissipate their energy so as to protect shorelines. See *wave breaker*.

bridge   A structure spanning a waterway that provides passage.

brine lake   See *brine lake* under *lake*.

broken flow   See *broken flow* under *flow*.

brook   Small, natural freshwater stream. See *creek*.

brushline   The edge of brushy cover near deep, more open water.

brushpile   See *brushpile* under *habitat enhancements*.

buffer   Vegetation strip maintained along a stream or lake to mitigate the impacts of actions on adjacent lands. Also called a buffer strip, leave strip, or streamside management zone.

bulk density   The ratio of the mass of bed material to its volume.

bulkhead   See *bulkhead* under *habitat enhancements*.

bund   Synonymous with *berm* under *habitat enhancements*.

buoy   A distinctively shaped and marked float, anchored to mark a channel, anchorage, navigational hazard, recreation area, or to provide a mooring place.

buoyancy   Ability of an object to float in water.

buried channel   (1) A channel that has been filled by unconsolidated deposits. (2) A water course that has been diverted through a pipe or culvert.

burn   Term sometimes used for a small stream.

burst (darting) speed   See *burst (darting) speed* under *swimming speed*.

butte   An isolated, usually flat-topped geological formation with a top width less than the height of the formation; formed by differential erosion of substrates.

▶ **C**

caldera lake   See *caldera lake* under *lake*.

caliche   Layer near the surface (or exposed by erosion) that is more or less cemented by carbonates of Calcium and Magnesium precipitated from the soil.

calving   (1) Erosion and collapse of sandbanks during a rapid drawdown of a reservoir, river, or lake that results from latent water storage. (2) Breaking away of a mass of ice from its parent glacier, iceberg, or sea-ice formation.

canal   An artificially created waterway.

canopy   Overhead cover of branches and foliage of adjacent vegetation.

canopy closure   The completeness of tree cover. See *canopy cover*.

canopy cover   Percentage of ground or water covered by shade from the outermost perimeter or natural spread of foliage from plants. Small openings within the canopy are excluded if the sky is visible through them. Total canopy coverage may exceed 100% due to the layering of different vegetative strata. See also *foliar cover* and *stream surface shading*.

canopy density   See *canopy cover*.

canopy layer   Foliage layer in a plant community.

canyon   A water-cut, deep chasm or gorge with steep sides, often with a stream at the bottom. Most characteristic of arid or semi-arid regions where downcutting by streams greatly exceeds weathering.

capacity   Maximum amount of water, sediment, or debris that can be carried by a stream.

capacity : inflow ratio   The storage capacity of a reservoir or lake expressed as a ratio to the mean annual inflow.

cape   A piece of land jutting into a large water body.

capillary   A tube or a fine bore channel. Channel in the material below the earth's surface through which water can move.

capillary fringe   A zone immediately above the water table in which water is drawn upward.

capillary percolation   Movement of water along a capillary channel.

capillary porosity   Portion of soil porosity that remains filled with water at a defined level of drainage.

capillary stream   Totality of water moving in a common direction along adjoining groundwater capillaries.

capillary water   Water stored or moving through groundwater capillary system.

capillary waves   See *capillary waves* under *waves*.

capture   See *stream capture*.

carp mumblings   Term used for small depressions about 0.64 cm (one-quarter inch) deep made by carp feeding in soft mud bottoms. Can be readily confused with small depressions that are made by other aquatic organisms.

carr   See *carr* under **wetlands**.

carrying capacity   The maximum biomass of aquatic organisms that can be sustained on a long-term basis by an aquatic ecosystem within existing environmental conditions. The maximum number (instead of biomass) is applied to terrestrial organisms.

cartesian   See *cartesian* under **remote sensing**.

cartesian well   Synonymous with **artesian well**.

cascade   See *cascade* under *turbulent—fast water* under the main heading **channel unit**.

cascading   The flow of water over one or a series of well-defined drops in stream.

catadromous   Life history strategy, that includes migration between fresh- and saltwater, in which fish reproduce and spend their early life stages in saltwater, move into freshwater to rear as subadults, and return to saltwater to spawn as adults. Compare with **anadromous, diadromous, potamodromous, oceanadromous**.

catastrophic drift   Term applied to the massive drift of bottom organisms that occurs under catastrophic or stress conditions such as floods or chemical toxicity.

catastrophic event   A large-scale, high-intensity natural disturbance that occurs infrequently.

catchment   The land area above a specified point from which water drains towards a stream, lake, reservoir, or sea.

catchment area   The total area draining into a given stream, lake, or reservoir. See **drainage area**.

catchment basin   The upslope land from which water drains by subsurface and surface routes into the lowest depression, especially for a reservoir or river. See **drainage basin**.

causeway   A raised road or path made across low water or wet ground.

cave-in lake   See *cave-in lake* under **lake**.

cell   See *cell* under **remote sensing**.

cell size   See *cell size* under **remote sensing**.

cemented bottom   Compacted or possibly cohesive substrate of a stream or lake, with particles adhering so tightly that penetration or erosion is difficult.

centripetal stream   See *centripetal stream* under **stream**.

channel   A natural or artificial waterway that periodically or continuously contains moving water, has a definite bed, and has banks that serve to confine water at low to moderate streamflows.

channel bank   The sloping land bordering a channel. Such a bank has steeper slopes than the bottom of the channel and is usually steeper than the land surrounding the channel.

channel billabong   See *channel billabong* under **billabong**.

channel bottom   The submerged portion of the channel cross section.

channel confluence pool   See *channel confluence pool* under *pool, scour pool* under the main heading **channel unit**.

channel constrictor   See *channel constrictor* under **habitat enhancements**.

channeled colluvium   See *channeled colluvium* under **valley segments**.

channel-forming discharge   Streamflow of a magnitude sufficient to mobilize significant amounts of the bed load.

channel geometry   The geomorphic form of the stream channel in the landscape surface.

> **aggraded channel**   A channel that has accumulated bed load, raising the elevation of the stream channel.
>
> **confined channel**   Well-defined channel of sufficient stability to remain in the same location and plane.
>
> **degraded channel**   A downcut channel that has scoured accumulated bed load materials.
>
> **incised channel**   Deep, well-defined channel with narrow width : depth ratio and limited or no lateral movement. Often newly formed, and is a result of rapid down-cutting into the substrate.

channelization   The mechanical alteration of a stream usually by deepening and straightening an existing stream channel or creating a new channel to facilitate the movement of water. See *confinement.*

channel junction bar See *channel junction bar* under *bar.*

channel maintenance or preservation flow The

minimum streamflow to sustain biota. See *channel maintenance or preservation flow* under *flow.*

channel pattern The configuration of a stream as seen from above and described in terms of its relative form, including:

abandoned meander channel Former stream channel that was cut off from a river and typically lacks standing water during the entire year.

anastomizing channel Multiple channels that diverge and converge around many islands.

apparently stable channel Condition of river channel where signs of lateral channel instability do not exist.

backbar channel Channel formed behind a bar that is connected to the main channel but usually at a higher streambed elevation than the main channel.

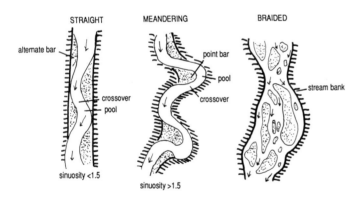

channel patterns (from Kohler and Hubert 1993)

braided channel Multiple channels, often with banks poorly defined.

dendritic channel Pattern of streamflow when tributaries progressively branch to form a tree-like pattern.

deranged channel The opposite of dendritic channels, irregular, poorly differentiated from the main channel and each other. They enter the main channel from different angles. Commonly found in recently glacial areas and large alluvial floodplains.

eddy return channel Small channel formed by recirculating flow of an eddy, usually between a river and a depositional bar. Also referred to as a backwater.

irregular channel Irregular sinuous channel; channel displays irregular turns and bends without repetition of similar features.

irregular meander channel A channel with no repeating pattern.

main channel Primary watercourse containing the major streamflow. Also referred to as main stem.

main stem channel See *main channel* under *channel pattern.*

meander scrolls channel Depressions or rises on the convex side of bends formed as the channel migrated laterally down valleys and toward the concave bank.

meltwater channel Channel formed by glacial meltwater.

regular meander channel A clear repeated meander pattern formed in a simple channel that is well-defined by cutting outside of a bend.

secondary channel Channel that flows laterally but parallel to the main channel, containing less water, and may be intermittent or completely dry during periods of low streamflow.

serpentine channel See *regular meander channel* under *channel pattern.*

side channel A secondary channel containing a portion of the streamflow from the main or primary channel.

sinuous channel Having a series of curves, bends, and turns. See *sinuosity.*

straight channel Very little curvature within the reach.

tortuous meander channel A repeated pattern characterized by angles greater than 90°.

wandering meander channel A stream channel that migrates in a random pattern across a floodplain.

channel sensitivity Capacity of a stream channel to respond to physical disturbance.

channel stability A measure of the resistance of a stream to changes in its unique form, channel dimensions, and patterns that determines how well it adjusts to and recovers from the changes in quantities of flow or sediment.

channel storage   Volume of water at a given time in the channel or over the floodplain of streams in a drainage basin or river reach.

channel type   A system for characterizing channels based on features such as channel and valley confinement, gradient, and erosional and depositional processes, such as:

Confinement	Gradient	Sediment process
Confined	>4%	Source
Moderately confined	1.5–4%	Transport
Unconfined	<1.5%	Response

Note: See Tables 1–3 for various systems for characterizing channel types.

channel unit   Relatively homogeneous areas of a channel that differ in depth, velocity, and substrate characteristics from adjoining areas, creating different habitat types in a stream channel.

fast water—turbulent   Channel with a gradient that exceeds 1%. A channel unit of this type possesses supercritical flow with hydraulic jumps sufficient to entrain air bubbles and create whitewater.

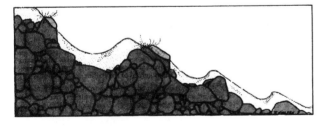

cascade (from Helm 1985)

cascade   Highly turbulent series of short falls and small scour basins, with very rapid

water movement as it passes over a steep channel bottom with gradients exceeding 8%. Most of the water surface is broken by short, irregular plunges creating whitewater, frequently characterized by very large substrates, and a well-defined stepped longitudinal profile that exceeds 50% in supercritical flow.

chute   Rapidly flowing water within narrow, steep slots of bedrock.

falls   Free-falling water with vertical or nearly vertical drops as it falls over an obstruction. Falling water is turbulent and appears white in color from trapped air bubbles.

  classic falls   Well-defined falls over a sheer drop.

  complex falls   Falls with a series of drops, breaks, or channels.

  curtain falls   Falls with a broad, uninterupted face.

  flume falls   Falls within a narrow, confining channel.

  ribbon falls   Elongate, narrow falls.

rapids (from Helm 1985)

rapids   Moderately steep stream area (4–8% gradient) with supercritical flow between 15 and 50%, rapid and turbulent water movement, surface with intermittent whitewater

TABLE 1.—Channel Types of Paustian et al. (1983)

Type	Hydrology–slope	Sediment	Valley form
A	Precipitation headwaters	Sediment source	Upper slopes steep
B	Precipitation transition zone	Sediment transport	Slope–valley interface
C	Precipitation runoff	Sediment deposition	Valley bottom
D	Glacial runoff	High sediment load	Glacial outwash valleys
E	Tidal	Sediment deposition	Estuarine

with breaking waves, coarse substrate, with exposed boulders at low flows, and a somewhat planar longitudinal profile.

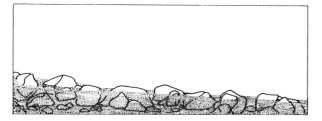

riffles (from Helm 1985)

riffles   Shallow reaches with low subcritical flow (1–4% gradient) in alluvial channels of finer particles that are unstable, characterized by small hydraulic jumps over rough bed material, causing small ripples, waves, and eddies, without breaking the surface tension. Stable riffles are important in maintaining the water level in the pool immediately upstream of the riffle.

> **high gradient riffle**   A collective term for rapids and cascades. Steeper reaches of moderately deep, swift (greater than 4% gradient), and very turbulent waters. Generally, these riffles have exposed substrates that are dominated by large boulders and rocks. See *cascade* and *rapids* under *fast water—turbulent* under the main heading ***channel unit***.

> **low gradient riffle**   Shallow reaches with swiftly flowing (gradients less than 4%), turbulent water with some partially exposed substrate, usually cobble or gravel.

step run   Low gradient runs with small (0.5 to 2 m) riffle steps between runs.

fast water—nonturbulent   Reaches that are deeper than riffles, with little or no supercritical flow. The water surface in such reaches has a smoother, laminar appearance.

> **run**   Swiftly flowing stream reach with a gradient greater than 4%, little to no surface agitation, waves, or turbulence, no major flow obstructions, approximately uniform flow, substrates of variable particle size, and water surface slope roughly parallel to the overall stream gradient.

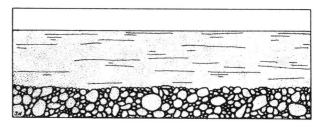

run (from Helm 1985)

> **sheet**   Shallow water reach that flows uniformly over smooth bedrock. Also referred to as a slipface.

slow water   Stream channel with a gradient of less than 1% that is typically deeper than the reach average with a streambed composed of finer substrates and a smooth, unbroken water surface.

> **edgewater**   A shallow, quiet area along the margins of a stream with water velocity that is low or nonexistent. Edgewater areas are typically associated with riffles.

> **embayment**   An off-channel, pond-like water body that has a connection (sometimes narrow) to the stream channel.

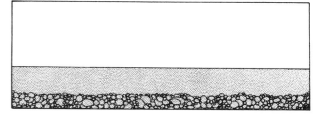

glide (from Helm 1985)

glide   A shallow stream reach with a maximum depth that is 5% or less of the average stream width, a water velocity less than 20 cm (8 in) per second, and without surface turbulence.

pool   Aquatic habitat in a stream with a gradient less than 1% that is normally deeper and wider than aquatic habitats immediately above and below it.

> **dammed pool**   Pool formed by impounded water from complete or nearly complete channel blockage caused by a beaver dam, log jam, rock slide, or stream habitat improvement structure. A dammed pool may form by substrate deposition at the confluence of a tribu-

TABLE 2.—Channel Types of Rosgen (1996)

Type	Description	Slope	Landform
Aa	Very steep, deeply entrenched with debris transport.	>10%	High relief, deeply entrenched and erosional. Vertical steps with deep scour pools and waterfalls.
A	Steep, entrenched, step-pool with high energy and debris transport.	4–10%	High relief, entrenched and confined. Cascading reaches with frequently spaced deep pools in a step-pool bed morphology.
B	Moderately entrenched, moderate gradient, riffle-dominated, infrequently spaced pools with very stable banks and profile.	2–3.9%	Moderate relief, colluvial deposition and (or) residual soil, moderate entrenchment, and moderate width : depth ratio. Predominately rapids with occasional pools in a narrow, gently sloping valley.
C	Low gradient, meandering, point bar-, riffle-, pool-, alluvial channels with broad, well-defined floodplain.	<2%	Broad valley with terraces associated with the floodplain, alluvial soils, slightly entrenched, and well-defined meandering channel. Riffle-pool streambed morphology.
D	Wide channel with longitudinal and transverse bars with eroding banks.	<4%	Broad valley with abundant sediment in alluvial and colluvial fans, glacial debris, and other depositional features exhibiting active lateral adjustment.
Da	Anastomosing channels that are narrow and deep with stable banks, very gentle relief, highly variable sinuosity, and an expansive well-vegetated floodplain and associated wetlands.	<0.5%	Broad, low-gradient valleys with fine alluvium and (or) lacustrine soil. Anastomosing geologic control creating fine deposition with well-vegetated bars that are laterally stable and a broad wetland floodplain.
E	Low gradient, riffle-pool with very efficient and stable meandering rate, low width : depth ratio, and little deposition.	<2%	Broad valley-meadow. High sinuosity with stable well-vegetated banks and floodplain of alluvial material. Riffle-pool morphology with very low width : depth ratio.
F	Entrenched meandering riffle-pool with a low gradient and high width : depth ratio.	<2%	Entrenched in highly weathered material with gentle gradient and a high width : depth ratio. Riffle-pool morphology with meandering channel that is laterally unstable with high bank erosion.
G	Entrenched "gully" step-pool with moderate gradient and low width : depth ratio.	2–3.9%	Gully, step-pool morphology with moderate slopes, low width : depth ratio, narrow valleys that are deeply incised alluvial or colluvial material. Unstable with grade control problems and high bank erosion rates.

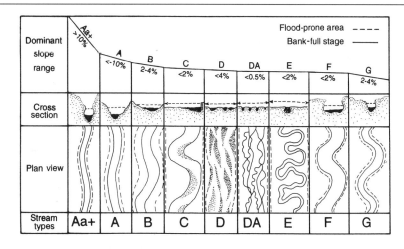

Channel type classification (from Rosgen 1994 with permission of Elsevier Science)

TABLE 3.—Channel Types of Montgomery and Buffington (1993)

Channel type	Description
Cascade	High gradient stream with large substrate where the flow is strongly three-dimensional and energy dissipation is dominated by tumbling jet-and-wake flow and hydraulic jumps.
Step-pool	A large series of steps created by larger substrate that separate pools with finer substrates.
Plane-bed	Lack of a well-defined bedform that is characterized by long reaches of relatively planar channel bed with occasional rapids.
Pool-riffle	Undulating channel bed with a sequence of bars, pools, and riffles.
Regime	Low-gradient, sandbed channel that exhibits a succession of bedforms with increasing flow.
Braided	Braided pattern of medial and longitudinal bars that are wide and shallow with a high sediment supply.

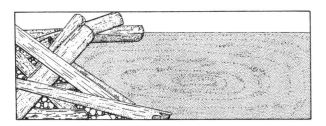

dammed pool (from Helm 1985)

beaver pool   Pool formed behind a dam created by beaver.

debris pool   Pool formed behind an a channel obstruction created by an accumulation of woody debris.

landslide pool   Pool created due to channel obstruction by materials transferred into the channel from adjacent slope or channel failures.

tary stream with the main stem river when water velocity decreases.

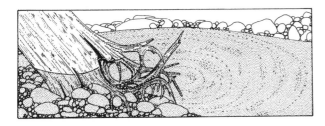

backwater (from Helm 1985)

backwater   (1) A pool formed by water backing upstream from an obstruction, such as narrowing of the channel by a bedrock or boulder constriction. (2) Abandoned channel that remains connected to the active main stem river. (3) Secondary channel in which the inlet becomes blocked with substrate deposition when water velocities decrease as the river subsides but the outlet remains connected with the active main channel.

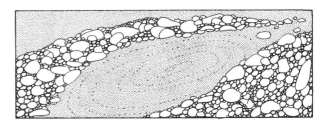

secondary channel (from Helm 1985)

secondary channel   Relatively small pools formed outside the mainstream wetted channel, sometimes separated by formation of a bar deposited along the margin of the main channel. Also referred to as an abandoned channel or side channel pool.

side channel   Elongated extension off the main channel that becomes a backwater under low streamflows when the inflow to the channel becomes blocked from sediment deposition.

slackwater pool   Pool-like depressions on the floodplain with beds of rock or coarse material and higher current velocities flowing in a uniform direction that contain water only during high flow or after floodwaters recede, more transient in nature than secondary channel pools, and may contain water for only a few days or weeks.

scour pool   Pool created by the scouring action of current flowing against an obstruction, causing an increase in lift and drag forces, a result of flow deflection, constriction, or increased local turbulence induced by a nonalluvial obstruction.

alcove (from Helm 1985)

alcove   A deeper area along the shoreline in a larger habitat where the stream is generally wide and shallow. Also referred to as a sidepool.

channel confluence pool   (1) The location where two streams converge. (2) A pool created by scour where two channels meet that has more turbulence and higher water velocities than found in many other types of pools. Also referred to as a channel convergence pool.

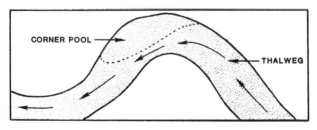

corner pool (from Helm 1985)

corner pool   Pool formed by lateral scour and transverse currents near the concave bank of a meander curve.

eddy   A pool on the margin or off the main channel of a stream that is formed and maintained by strong eddy currents.

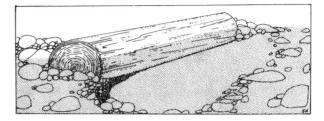

lateral scour pool (from Helm 1985)

lateral scour pool   A pool formed by the scouring action of the flow as it is directed laterally or obliquely to one side of the stream by the configuration of the channel or a partial channel obstruction. Usually confined to less than 60% of the channel width.

main channel pool   A pool covering the entire channel; typically associated with one bedrock bank and a bend in the stream.

mid-channel pool   A large pool formed by mid-channel scour that encompasses greater than 60% of the wetted channel with low velocity.

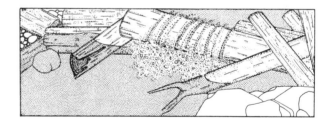

plunge pool (from Helm 1985)

plunge pool   A pool created by water passing over or through a complete or nearly complete channel obstruction, and dropping steeply into the streambed below scouring out a basin in the stream substrate where the flow radiates from the point of water entry. This is an example of hydraulic control in a stream that determines how energy of moving water shapes a channel. Also referred to as a falls pool or plunge basin.

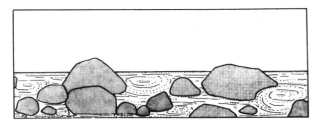

pocket water (from Helm 1985)

underscour pool (from Helm 1985)

pocket water   One or a series of small pools in a section of swiftly flowing water containing numerous obstructions such as boulders or logs which create eddies or scour holes in the channel substrate. Typically found in cascades and rapids.

residual pool   The pool portion that lies below the elevation of the downstream outlet crest.

straight   Straight, elongated pool in the center of a channel that is created by an upstream constriction.

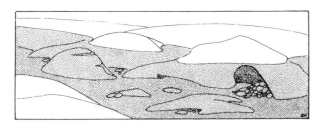

trench (from Helm 1985)

trench   A relatively long, deep slot-like depression in the streambed, typically with a U-shaped channel, coarse-grained or bedrock substrates, and high water velocities. Such reaches are usually found in highly confined, often bedrock-dominated channels.

underscour pool   Form of scour pool created by a log or other obstruction near the surface that causes the water to be deflected downward, scouring out a pool in the substrate. Also referred to as an upsurge pool.

channel width   See *channel width* under **dimensions**.

charco   See *charco* under **pond**.

check dam   See *check dam* under **habitat enhancements**.

chemical oxygen demand (COD)   Represents the reduction capacity of organic and inorganic matter present, and is the amount of molecular oxygen required to stabilize the proportion of the sample which is susceptible to oxidation by a strong chemical oxidant.

chemical stratification   Layering of water in a lake because of density differences or differential concentrations of dissolved substances with depth.

chemocline   See *chemocline* under **stratification**.

chronic toxicity   Sublethal toxicity of long duration that adversely affects an organism and may eventually lead to death. The sublethal toxicity is reflected through changes in productivity and population structure of a community.

chunk rock   Formation of mixed sized, irregular-shaped rocks caused by gravity or erosion, usually found at the base of a bluff or cliff.

chute (from Helm 1985)

chute   (1) A narrow, confined channel through which water flows rapidly. (2) A rapid or quick descent in a stream, usually with a bedrock substrate. (3) A short straight channel which bypasses a long bend in a stream, and is formed by the stream cutting across a narrow land area between two adjacent bends. See also *chute* under *fast water—turbulent* under **channel unit**.

cienagas   See *cienagas* under **wetlands**.

cirque   Rounded, bowl-like depressions in mountains that are created by weathering, erosion, and glacial action.

cirque lake   See *cirque lake* under **lake**.

class   See *class* under **remote sensing**.

classic falls   See *classic falls* under *fast water—turbulent, falls* under the main heading **channel unit**.

clast   An individual particle, detrital sediment or a sedimentary rock, initially produced by the disintegration of a larger mass of bedrock, classified according to size.

clay   Natural earthy material which is plastic when wet, and consists essentially of hydrated silicates of aluminum, less than 4 μm. Compare with other substrate sizes under **substrate size**.

clay bottom   Bottom composed of clay or clay-like material.

cleanwater association   An association of organisms found in any natural, unpolluted environment that is characterized by species sensitive to environmental changes caused by contaminants or pollutants.

cliff   The high steep face of a rocky mass overlooking a lower area; a precipice. See **bluff**.

climate   The meteorological conditions including temperature, precipitation, wind, pressure, evaporation, and transpiration that prevail in and characterize a location.

climatic year   A continuous 12-month period during which a complete annual cycle occurs. The U.S. Geological Survey uses the period October 1 through September 30 in the publication of its records of streamflow as a water year. See **water year**.

climax succession stage   See *climax succession stage* under **succession**.

clinograde   See *clinograde* under **stratification**.

clinolimnion   See *clinolimnion* under **stratification**.

closed basin   A basin without a surface outlet, from which water is lost only by evaporation or percolation.

closed lake   See *closed lake* under **lake**.

clump   A relatively dense aggregation of vegetation of the same species. See also *clump* under **large organic debris** and **remote sensing**.

cluster   See *cluster* under **remote sensing**.

coarse load   See *coarse load* under **sediment load**.

coarse particulate organic matter   See *coarse particulate organic matter (CPOM)* under **organic particles**.

coarse woody debris   See **large organic debris**; also a synonym for **large woody debris**.

coast   Land next to the sea.

coastal delta floodplain   See *coastal delta floodplain* under **floodplain**.

coastal lake   See *coastal lake* under **lake**.

coastal plain   A relative flat area, or plain, extending along a coast.

coastline   Shape or pattern of an ocean coast.

cobble   Stream substrate particles between 64 and 128 mm (2.5–5 in) in diameter. Compare with other substrate sizes under **substrate size**.

coefficient of storage   See *coefficient of storage* under **groundwater**.

coffer dam   Temporary structure constructed in a water body to provide a dewatered area for construction of structures such as bridges.

cohesion   State in which particles of a single substance are held together by primary or secondary valence forces.

col   A pronounced dip in a ridge, or between two peaks, connecting a neck of land; often formed at a divide where water courses flow in opposite directions.

cold monomictic   See *cold monomictic* under **mixing**.

cold spring   See *cold spring* under **spring**.

coldwater fishes   A broad term applied to fish species that inhabit waters with relatively cold temperatures (optimum temperatures generally between 4–15°C (40–60°F). Examples are salmon, trout, chars, and whitefish. Compare with *coolwater fishes, warmwater fishes*.

coldwater lake   See *coldwater lake* under *lake*.

collectors   See *collectors* under *macroinvertebrates*.

colloidal   Particles which are 10^{-7} to 5×10^{-6} mm in diameter, larger than most inorganic molecules, and that usually remain suspended indefinitely in the water column.

colluvial   Gently inclined surface at the base of a slope that represents a transition zone between landforms. Colluvium is characterized by erosion and transport, and downslope sites of deposition. See also *colluvial* under *streambank material*.

colluvial soil   Recently transported soil derived from material eroded and deposited locally through sheet flow. In its extreme form, such as avalanche or landslide, the soil comprises material of many sizes.

colluvium   A general term for loose deposits of soil and rock moved by gravity (e.g., talus). See also *colluvium* under *valley segments*.

colonization   The establishment of a species in an area not previously occupied by that species.

color   The quality of water with respect to reflected or refracted light, measured as a wavelength pattern or hue.

community   An assemblage of plants and animals occupying a given area; two or more populations of organisms interacting within a defined time and space.

community tolerance quotient (CTQ)   See *community tolerance quotient (CTQ)* under *biological indices*.

compaction   An increase in the density of a material by reducing the voids between the particles. The relative density of bed material, usually caused by sedimentation, mineralization, or imbrication.

compensation   Creation or restoration of wetland areas that are equivalent to areas and functions of destroyed wetlands.

compensation level   Depth in a water body at which the available light is reduced enough to cause photosynthesis to equal respiration. Also referred to as the compensation point.

competence   Maximum size of particle that a stream can carry, which depends on water velocity and gradient. Also defined as the critical stress necessary for grain movement.

complex falls   See *complex falls* under *fast water—turbulent, falls* under the main heading **channel unit**.

complexity   Term used to describe the presence of a variety of habitat types within a defined area of a waterbody. Increased complexity provides habitat for a greater variety of organisms or life stages, and is usually an indicator of better habitat health.

compound meander   See *compound meander* under *meander*.

concave bank   See *concave bank* under **streambank**.

concretion   A localized concentration of chemical compounds (e.g., calcium carbonate and iron oxide) in the form of a soil grain or nodule of varying size, shape, hardness, and color; concretions of significance in hydric soils are usually iron oxides and manganese oxides occurring at or near the soil surface, the result of fluctuating water tables.

conductivity   A measure of the ability of a solution to carry an electrical current. Conductivity is dependent on the total concentration of ionized substances dissolved in the water and is measured as microsiemens per centimeter.

conduit   A pipe, tube, or the structure for conveying water or other fluids.

confined   See *confined* under *confinement*.

confined aquifer   An aquifer that is restricted in size by impervious materials.

confined channel   See *confined channel* under *channel geometry*.

confined meander   See *confined meander* under *meander*.

confinement   Degree to which the river channel is limited in its lateral movement by valley walls or relic terraces.

channelization   Deepening an existing stream channel or creating a new stream channel by human activity to increase the rate of runoff or to lower the water table.

confined channel   A stream that is in continuous or repeated contact at the outside of major meander bends.

entrenched channel   A stream bend that is in continuous contact with bedrock valley walls or terraces.

frequently confined channel   A stream that is frequently confined by the valley walls or terraces.

secondarily confined channel   A stream channel that has cut down into deposited sediments because a controlling structure has been lost.

unconfined channel   A stream channel that is not touching the valley wall or terrace and is capable of lateral migration.

confluence   The location where two streams flow together to form one.

conglomerate   Cemented material; rock consisting of rounded and waterworn gravel imbedded in a finer material.

conjugate points (conjugate principle points)   See *conjugate points (conjugate principle points)* under *remote sensing*.

connate water   Water trapped in deep geological sediments at the time the sediment was deposited.

connectivity   Water exchange between the river channel and the associated floodplain.

consequent stream   See *consequent stream* under *stream*.

conservation pool   The minimum water level that is normally reserved behind a dam for a variety of purposes.

conservation storage   Storage of water for later release for useful purposes (e.g., municipal water supply, power, or irrigation) in contrast with storage of water used for flood control.

constriction   (1) A reduction in the channel width by a resistant structure. (2) Location where the river channel is prevented from migrating laterally within the valley, usually by a bedrock ridge protruding from the valley wall.

consumptive water use   Occurs when water is removed and not returned.

contamination   Presence of elevated levels of compounds, elements, physical parameters, or substances that make the water impure or unsuitable for use.

contents   Volume of water in a reservoir. Unless otherwise indicated, reservoir content is computed on the basis of a level pool and does not include bank storage.

contiguous habitat   Habitat able to provide the life needs of a species that is distributed in a continuous or nearly continuous pattern.

continuous gully   See *continuous gully* under *gully*.

continuous stream   See *continuous stream* under *stream*.

contour   An imaginary line of constant elevation related to the surface of the earth.

controlling breaching   See *controlling breaching* under *breaching*.

control station   Any streamflow measurement site where a regulatory base flow has been established.

control structure   Artificial structure designed to regulate and control the movement of water.

draft tube   A conduit extended from a turbine.

drop inlet   A water level control structure with a vertical tube connected to a horizontal tube that discharges through a dike or dam.

drum gate   A circular gate at the entrance of a spillway.

gallery   Area of a water intake structure behind a trash rack where water is distributed across the face of a screen to prevent fish and small debris from entering the water system.

gate   Moveable structures used to control the movement, storage, and drainage of water.

flap gate   Moveable gate at a right angle to a pipe that responds to a change in flow. The current opens the gate to drain and pushes the gate against the pipe to prevent re-entry of water.

radial gate   A gate on pivotal arms.

roller gate   Similar to a sliding gate but operating on rollers.

sliding gate   Horizontal gate that is hoisted up or down between guides.

high tube overflow   Horizontal tube on a small impoundment that can accommodate spill.

natural spillway   Natural, undisturbed ground sufficiently firm to handle overflow from an impoundment.

ogee   Spillway with an S-shaped weir.

outlet tower   Structure that provides water control to a powerhouse inlet.

penstock   Intake for a turbine.

ski jimp   Spillway shaped like a ski jump.

spillway   Part of a control structure designed to allow water to spill over a dam without weakening it.

stop log   Box, logs, or planks with a drain pipe at the bottom that extends through the wall of an impoundment.

tin whistle   Vertical riser tube connected to a horizontal tube through the wall of an impoundment. Stop logs are placed around the riser so that the water level is higher than the top of the riser allowing water to spill over the logs into the vertical tube.

trash rack   Protective structure used to intercept logs and other debris.

turbine   Rotary motor for converting the energy of falling water into electricity.

convex bank   See *convex bank* under **streambank**.

conveyance loss   Loss of water in transit from a conduit or channel due to leakage, seepage, evaporation, or evapo-transpiration.

convolution   The sinuosity of a stream channel. The convolution index is the same as the sinuosity index.

cooling pond   See *cooling pond* under **pond**.

cooling water   Water used to cool operating equipment of a power plant or other industrial facility.

coolwater fishes   A broad term applied to fish species that inhabit waters with relatively cool temperatures (optimum temperatures generally between 10 and 21°C (50 and 70°F). Compare with *coldwater fishes, warmwater fishes*.

copropel   See *gyttja*.

coriolis effect   (1) Force deflecting water currents as a result of the earth's rotation. The deflection is to the right in the Northern Hemisphere and to the left in the Southern Hemisphere. (2) A force acting at right angles to horizontal path of particles in motion.

corner pool   See *corner pool* under *slow water, pool, scour pool* under the main heading **channel unit**.

corrasion   (1) The wearing down of underwater surfaces by chemical solutions. (2) The lateral and vertical cutting action of water courses, through abrasive power of their loads. (3) The wearing down of underwater surfaces through natural friction, as in the moving bed load of rivers. Sometimes referred to as abrasion.

counter dam   A small wall built across the apron of the foot of a key dam to prevent it from being undermined by erosion.

course   The direction of flow of water across the earth's surface.

cove   A small indentation or recess in the shoreline of an aquatic system, sea, lake, or river; a sheltered area. See *arm, bay, embayment*.

cover   Structural materials (boulders, logs, or stumps), channel features (ledges, vegetation), and water features (turbulence or depth) that provide protection for aquatic species. See also *cover* under **habitat enhancements**.

crater lake   See *caldera lake* and *volcanic lake* under *lake*.

creek   A small lotic system that serves as the natural drainage course for a small drainage basin. See **brook**.

creep   Gradually and generally imperceptible downhill movement of soil and loose rock; a slow moving type of landslide.

creeping flow See *creeping flow* under *flow*.

crenocole An organism living only in spring environments.

crenogenic meromixis See *crenogenic meromixis* under *mixing*.

crenophile An organism that prefers spring environments but may be found in similar types of habitat.

crest (1) Top of a dike, spillway, or weir to which water must rise to pass over. (2) The summit or highest point of a wave. (3) Highest elevation reached by flood waters flowing in a channel.

crib See *crib* under *habitat enhancements*.

cribwell See *cribwell* under *habitat enhancements*.

critical depth Minimum depth that can occur as water flows over the top of a boulder, log, or crest of a spillway. Occurs slightly upstream of the brink.

critical flow State of water flow when one or more properties such as velocity undergoes a change.

critical location Location along a stream with a minimum concentration of dissolved oxygen.

critical reach See *critical reach* under *reach*.

critical slope Slope that sustains a uniform critical flow.

critical velocity See *critical velocity* under *velocity*. Also referred to as *critical flow*.

cross-ditch A ditch excavated across the road at an angle and depth sufficient to divert both road surface water and ditch water off or across the road.

cross-over bar See *cross-over bar* and *transverse bar* under *bar*. Definition is provided with *transverse bar* under *bar*.

cross-sectional area See *cross-sectional area* under *dimensions*.

crown The growing tip of a tree.

cryogenic lake See *cryogenic lake* under *lake*.

cryptodepression Portion of a lake below sea level.

cuesta Coastal plain ridge that is steepest toward the continent, due to the differential erosion of sedimentary deposits that gently dip seaward.

cultural eutrophication Accelerated addition of nutrients to a water body by human activities.

culvert A passage, usually a pipe, constructed beneath a road, railroad, or canal to transport water.

current Water moving continuously in one direction; the speed at which water is moving. See *velocity*.

curtain falls See *curtain falls* under *fast water—turbulent, falls* under the main heading **channel unit**.

cusps Triangular deposits of sand, or other current drift, spaced along a shore.

cut bank See *cut bank* under **streambank**.

cutoff lake See *cutoff lake* under *lake*.

cutoff wall A wall (usually concrete) installed downstream of culverts to keep water and materials in place.

cutting back Upslope movement of a stream channel due to erosion. See also **head cut**.

cycle of erosion progressive stages in erosion of a landscape.

▶ d

D_{50} See d_{50} or D_{50} under *sediment load*.

daily discharge See *daily discharge* under *discharge*.

dam A barrier obstructing the flow of water that increases the water surface elevation upstream of the barrier. Usually built for water storage or to increase the hydraulic head.

damaging flood A flood of magnitude exceeding the normal maximum discharge.

dam-break flooding See *dam-break flooding* under *flooding*.

dammed pool See *dammed pool* under *slow water, pool* under the main heading **channel unit**.

Darcy's law See *Darcy's law* under *groundwater*.

deadhead A log floating at or near the surface that presents a hazard to boating. See also *deadhead* under *large organic debris*.

dead lake See *dead lake* under *lake*.

deadman A structure of wood, metal, or concrete that is buried in the ground to serve as an anchor for cables or other lines.

dead storage The volume in a reservoir below the lowest controllable level.

dead water Water without measurable currents. Generally refers to water in a stream behind an obstruction.

debris Any material, organic or inorganic, floating or submerged, moved by water. Geologists and hydrologists have used this term in reference to inorganic material; more recently fishery workers have used the term in reference to organic material.

debris avalanche See *debris avalanche* under *landslide*.

debris fall See *debris fall* under *landslide*.

debris flow See *debris flow* under *landslide*.

debris jam or dam See *debris jam or dam* under *landslide*.

debris loading The quantity of debris located within a specific reach of stream channel due to natural processes or human activities.

debris pool See *debris pool* under *slow water*, *dammed pool* under *channel unit*.

debris slide See *debris slide* under *landslide*.

debris torrent See *debris torrent* under *landslide*.

deck Walking or work surface on an impoundment, dock, or boat.

deep-seated creep See *deep-seated creep* under *landslide*.

deep-seated failures See *deep-seated failures* under *landslide*.

deepwater habitat Permanently flooded lands lying below the deepwater boundary of wetlands. This boundary is 2 m (6.6 ft) below low water or at the edge of emergent macrophytes, whichever is deeper. Any open water in which the mean water depth exceeds 2 m (6.6 ft) at mean low water in nontidal and freshwater tidal areas, or below extreme low water at spring tides in salt and brackish tidal areas, or the maximum depth of emergent vegetation, whichever is greater.

deepwater zone An area of fairly great depth.

deflector See *deflector* under *habitat enhancements*.

degradation (1) Geologic process by which streambeds and floodplains are lowered in elevation by the removal of material. (2) A decline in the viability of ecosystem functions and processes. Compare with *aggradation*.

degraded channel See *degraded channel* under *channel geometry*.

deliverability Likelihood that, as a result of one or more land-use practices or through cumulative effects, a given amount of wood, water, sediment, or energy will be delivered to fish habitat in streams.

delivered hazard Adverse changes in the amount or location of wood, water, sediment, or energy being delivered downstream that may affect fish, water quality, or capital improvements.

delta Flat plane of alluvial deposits between the branches at the mouth of a river, stream, or creek. See *alluvial fan*.

delta bar See *delta bar* under *bar*.

delta kame Deposit with the form of a steep, flat-topped hill located at the front of a retreating continental glacier.

delta lake See *delta lake* under *lake*.

dendritic channel See *dendritic channel* under *channel pattern*.

density (1) Number of individuals per unit of surface area or volume. (2) Mass per unit volume.

density current A flow of water maintained by gravity through a large body of water, such as a reservoir or lake, that retains its unmixed identity because of density differences.

density stratification   See *stratification*.

deposit   An accumulation of organic or inorganic material resulting from naturally occurring biological, chemical, or physical processes.

depositing substrate   Bottom areas where solids are being actively deposited, often in the vicinity of effluent discharges.

deposition   Settling of material from the water column and accumulation on the streambank or bed. Occurs when the energy of flowing water is unable to support the load of suspended material.

deposition zone   Location along an erosion transport network where materials are deposited because water velocity and volume are insufficient to retain the materials in suspension.

depression   Any relatively sunken part of the earth's surface, especially low-lying areas surrounded by higher ground, that may be natural or constructed.

depression storage   Volume of water contained in natural depressions on the land surface (e.g., puddles).

depth   See *depth* under *dimensions*.

depth : area ratio   See *depth : area ratio* under *dimensions*.

depth integration   See *depth integration* under *sediment load*.

depth of scour   A relationship of the depth of scouring to the bed load transport rate:

$$d_s = \frac{Q_b}{u_b W p_s (1 - p)};$$

d_s = depth of scour;
u_b = average bed load velocity;
Q_b = bed load transport rate;
W = width of stream;
p = particle size.

deranged channel   See *deranged channel* under *channel pattern*.

desalinization   Removal of salts from brackish or marine waters.

desert river   See *desert river* under *river*.

desiccation   Process of dehydration or drying up.

design high water level   Elevation of the enveloped profile of the 50-year flood, or flood series, routed through the reservoir with a full conservation pool after 50 years of sediment, or the elevation of the top of the flood control pond, whichever is higher.

destratification   See *destratification* under *stratification*.

detention reservoir   An ungated reservoir for temporary storage of flood water.

detention storage   Volume of water, other than depression storage, flowing on the land surface and that has not yet reached the channel.

detrital sedimentation   Deposition of organic sediments.

detritus   A nondissolved product of disintegration or wearing away. Pertains to small organic particles like leaves and twigs. Detritus may pertain to material produced by erosion, such as soil, sand, clay, gravel, and rock, carried down a watercourse, and deposited on an outwash fan or floodplain.

dewatering   Removal of water from a site.

diadromous   Life history strategy, that includes movement between fresh- and saltwater, where organisms exhibit two migrations to spend various life stages in different ecosystems. Compare with *anadromous, catadromous, oceandromous, potamodromous*.

diagonal bar   See *diagonal bar* under *bar*.

diamond bar   See *diamond bar* under *bar*.

diamond-braided bar   See *diamond-braided bar* under *bar*.

diatom   Microscopic algae with a silaceous skeleton that occurs as plankton or attaches to substrate.

diel   Pertaining to a 24-hour period or a regular occurrence in every 24-hour period. See *diurnal*.

diffuser   A structure at or near the end of an outfall designed to improve the initial dilution, dispersal, or mixing of discharged effluent.

diffusivity   Rate at which a substance or temperature change will be transmitted through the water.

digger log   See *digger log* under **habitat enhancements**. Compare with *digger log* under **large organic debris**.

digital classification   See *digital classification* under **remote sensing**.

digital enhancement   See *digital enhancement* under **remote sensing**.

digital terrain model   See *digital terrain model* under **remote sensing**.

digitizing   See *digitizing* under **remote sensing**.

dike   An embankment for controlling water. Also used for any impoundment structure that completely spans a navigable water. Compare with *berm* under **habitat enhancements**.

dilution   Reduction in the concentration, or strength, of a substance by increasing the proportion of water in the mixture.

dimensions   The linear, areal, and volumetric features of an aquatic habitat as measured on both the horizontal and vertical scale.

    acre-foot   A unit for measuring the volume of water, equal to the quantity of water required to cover a surface area of one acre (0.4047 ha) to a depth of one foot (0.3048 m) and equal to 43,560 cubic feet, 325,851 gallons, or 1,233 cubic meters.

    area   Quantitative measurement of the surface of a body of water or drainage basin.

    average depth   Total of all depth measurements divided by the number of measurements taken at a site on a water body.

    average width   Total of all width measurements divided by the number of measurements taken at a site on a water body.

    bank depth   Vertical distance between the water surface and the floodplain.

    bank width   See *channel width* under **dimensions**.

    basin slope   Change in depth per unit of horizontal distance.

    breadth   See *width* under **dimensions**.

    channel width   Horizontal distance along a transect line from bank to bank at the bankful stage, measured at right angles to the direction of flow. Multiple channel widths are summed to represent total channel width.

    cross-sectional area   Area formed by the width and depth of a stream, channel, or waterway, measured perpendicular to the center line of flow.

    depth   Dimension of a water body measured vertically from the surface to the bottom.

    depth : area ratio   Area of a lake, by depth classification, expressed as a ratio to the total lake area.

    drainage basin shape (R_f)   Ratio of a basin's area (A) to the square of its maximum length (L):

$$R_f = A/L^2 \, .$$

    hydraulic depth (D)   Ratio of the cross-sectional area (A) divided by the width (W):

$$D = A/W \, .$$

    hydraulic radius (R)   Ratio of the cross-sectional area (A) divided by the wetted perimeter (P):

$$R = A/P \, .$$

    lake volume   An estimation of the total volume of a lake:

$$V = (h/3) \, (A_1 + A_2 + [A_1 A_2]^{\frac{1}{2}}) \, ;$$

    h = vertical depth;
    A_1 = area of the upper surface;
    A_2 = Area of lower surface.

    lake width   Distance on the lake surface from shore to shore at right angles to the length.

    length   Distance along the thalweg of a stream channel or the longest straight axis between the shores of a lake.

    maximum bank height   Maximum vertical distance from the water surface to the top of the highest bank.

    maximum depth   Greatest depth of the body of water. In streams, the greatest water depth at the sample location.

    maximum elevation   Highest point in the sample area, or the highest point in a watershed.

    maximum lake length   Single line distance on the lake surface between the most distant points on the lake shore.

maximum width   Greatest measurement from shore to shore along a line perpendicular to the thalweg of a stream or the greatest measurement across other bodies of water.

mean lake depth (Z)   Volume (*V*) divided by the surface area (*A*):

$$Z = V/A .$$

Compare with *average depth* under **dimensions**.

mean lake width   Quotient of the lake area divided by the maximum length.

mean stream length   Length of a stream segment along a line extending from the point of origin in a stream along the center of the channel to the point where the measurement is terminated.

median depth   Midpoint elevation between the maximum depth and the surface on a line from shore to shore.

median width   Midpoint width between the start of measurements and the maximum measurements.

minimum bank height   Minimum vertical distance from the water surface to the top of the lowest bank.

pool : riffle ratio   Ratio of the surface area or length of pools to the surface area or length of riffles in a given stream reach, frequently expressed as the relative percentage of each category.

pressure head   Relative pressure (excess over atmospheric pressure) divided by the unit weight of water, expressed in units of height.

radius of curvature (r_c)   Radius of the curve that describes the symmetrical meander of a stream or river. Is often used to evaluate channel resistance to erosion or the migration rates of bends and meanders.

$$r_c = \frac{L_m K^{-1.5}}{13\,(K-1)^{\frac{1}{2}}} ;$$

L_m = meander wavelength;
K = sinuosity.

relative depth (Z_r)   Maximum depth (Z_m) as a percentage of the mean diameter of a lake. With diameter expressed in terms of lake area (*A*), the formula is:

$$Z_r = \frac{88.6 \times Z_m}{A^{\frac{1}{2}}} .$$

relief ratio (R_r)   Ratio of a basin's length and altitude change in the basin:

$$R_r = h/L ;$$

h = difference in elevation between the river mouth and the highest point in the drainage
L = maximum length of a basin

shoreline : area ratio   Ratio of lake shoreline length to area of lake surface.

shoreline development (D_L)   Ratio of a lake's shoreline length (*L*) to the circumference of a circle having the same area (*A*) as the lake:

$$D_L = \frac{L}{2(\pi A)^{\frac{1}{2}}} .$$

shoreline length   Length of the perimeter of a lake.

sinuosity   An index (*K*) of a stream's meander as a function of stream length or valley profile.

$$K = \frac{L_c}{L_v} = \frac{S_v}{S_c} ;$$

L_c = channel length;
L_v = valley length;
S_v = valley slope;
S_c = channel gradient.

stream width   Distance between the two margins of flowing water in a stream at right angles to the flow.

top width   Width of a stream at the water surface that varies with changes in flow.

valley floor width index (VFWI)   Measure, in channel widths, of the variation in the width of a valley floor.

$$\text{VFWI} = \frac{(W_{ac} + W_{fp})}{W_{ac^*}} ;$$

W_{ac} = width of active channel;
W_{fp} = floodplain width;
W_{ac^*} = average width of active channel for all reaches at each site.

volume development   Comparison of the shape of a lake to the shape of an inverted cone equal to the lake's surface area.

wetted cross section   Total cross-sectional area through which a river flows above the bed at a specific discharge.

wetted perimeter   Length of the wetted contact between a stream of flowing water and the stream bottom in a vertical plane at right angles to the direction of flow.

wetted width   Width of a water surface measured perpendicular to the direction of flow at a specific discharge. Widths of multiple channels are summed to represent the total wetted width.

width   Measure of the cross section shape of a stream channel or across the narrow dimension of a lake, pond, or reservoir.

width : depth ratio   An index of the cross section shape of a stream channel, at bankful level.

dimictic   See *dimictic* under **mixing**.

dip   A hollow or depression in a land form.

dipslope   Conformation of land surface features with the shape of underlying bedded materials.

direct solar radiation   See *direct solar radiation* under **solar radiation**.

direct toxicity   Toxicity that has a direct—rather than indirect—effect on organisms, for example, chronic exposure to toxicants from contaminated food organisms.

discharge   (1) Rate at which a volume of water flows past a point per unit of time, usually expressed as cubic meters per second or cubic feet per second. (2) Intentional or unintentional release of substances into a waterway or water body that can occur from spilling, leaking, pumping, pouring, or dumping. (3) Any addition of dredged or fill material into a waterway or water body.

annual maximum daily discharge   Highest total daily discharge during the year.

annual maximum instantaneous discharge   Highest discharge recorded in a year.

annual mean discharge   Sum of total daily discharges for a year divided by the number of days in a year. Also, the total annual discharge divided by the number of seconds in a year.

annual minimum daily discharge   Lowest discharge recorded for any single day during a one-year period.

daily discharge   Total discharge from midnight to midnight for a continuous recording.

hydraulic discharge   An estimate of water volume (Q) passing by a point on a stream or river:

$$Q = WDu \, ;$$

W = width;
D = depth;
u = velocity of flow.

See *stream discharge* under **discharge** for an alternate meaning of Q.

instantaneous discharge   Discharge at a particular point in time.

mean monthly discharge   Average volume of water discharged per month during the given year. Expressed as hectares per meter.

monthly mean discharge   Average volume of water discharged per day during the given month.

stream discharge   An estimate of water volume (Q) passing a point on a stream or river:

$$Q = AVn \, ;$$

A = cross section area of stream channel;
V = velocity of water;
n = Manning's bed roughness constant.

See *hydraulic discharge* under **discharge** for an alternative meaning of Q.

discharge area   An area where water is released into surface water, groundwater, or the atmosphere.

disclimax   Term applied to a situation where recurring disturbances, such as grazing or periodic burning, exert a predominant influence in maintaining the species composition and density of vegetation.

discontinuity layer   See *discontinuity layer* under **stratification**.

discontinuous gully   See *discontinuous gully* under **gully**.

dished out bank   Streambanks that have a bank angle greater than 90°. See **streambank**.

dispersal   Multi-directional spread, at any time scale, of plants or animals from any point of origin to another, resulting in occupancy of other areas in their geographic range. Differs from emigration or immigration where movements are one-way from or into an area, respectively.

dispersion   (1) Separation or scattering of particles in water. (2) Active movement of individual organisms into adjoining areas.

dissolved load   Quantity of material dissolved in a specified volume of liquid.

dissolved organic matter   See *dissolved organic matter (DOM)* or *dissolved organic carbon (DOC)* under *organic particles*.

dissolved oxygen   Concentration of oxygen dissolved in water, where saturation is the maximum amount of oxygen that can theoretically be dissolved in water at a given altitude and temperature. Expressed as milligrams per liter or as percent saturation.

dissolved solids   Total of disintegrated organic and inorganic material that is dissolved in water. See *total dissolved solids*.

distorted meander   See *distorted meander* under *meander*.

distributary   See *distributary* under *stream*.

distribution   Occurrence, frequency of occurrence, position, or arrangement of animals or plants within an area. May also be applied to a rate such as the number per unit of area or unit of time.

distrophic (dystrophic)   See *distrophic* under *trophic*.

disturbance   A force that causes changes in habitat or community structure and composition through (a) natural events such as fire, flood, wind, or earthquake; (b) mortality due to insect or disease outbreaks; or (c) human activities such as agriculture, grazing, logging, mining, road construction, etc.

ditch   A long, narrow excavation in the ground (usually an open and unpaved channel, trench, or waterway smaller than a canal) for conveying water to or from a specific location for purposes such as drainage or irrigation.

diurnal   (1) Refers to events, processes, or changes that occur every day. (2) May be applied to organisms that are active during the day. See *diel*.

diversion   An artificial structure, such as a canal, embankment, channel, pipe, or other conduit, for taking water from a stream or other body of water to another location.

diversity   Variation that occurs in plant and animal taxa (i.e., species composition), habitats, or ecosystems within a given geographic location.

diversity index   Any of several numerical measures of animal or plant diversity in a given area. Often, the relationship of the number of taxa (richness) to the relative number of individuals per taxon (evenness) for a given community or area. See *diversity index* under *biological indices*. Also, see *habitat quality indices*.

divide   Topographical boundary in a catchment area that is related to the highest elevation surrounding the area. The hydrological divide would include subsurface as well as surface runoff. See *drainage divide*.

dock   A narrow walk or platform supported on posts or floats extending from the shore into a water body for the purpose of anchoring, docking, and landing boats, and for loading equipment and people into boats.

doline lake   See *doline lake* under *lake*.

domain   Region or area that is characterized by a specific feature, such as the type of vegetation, aquatic system, or wildlife.

domestic water supply   Water from wells, streams, lakes, or reservoirs used for human consumption and other uses in homes.

dominance and taxa (DAT)   See *dominance and taxa (DAT)* under *biological indices*.

dominant discharge   Stream discharge level(s), usually the bank-full flow, that in aggregate is sustained over a long enough time period to form and maintain a relative equilibrium in a natural channel by dislodging, transporting, and distributing bed materials. Also referred to as a formative discharge.

downcutting   Water erosion that deepens an existing channel or forms a new channel where one did not exist previously.

down log   See *down log* under *large organic debris*.

downstream link   See *downstream link* under *link*.

draft tube   See *draft tube* under *control structure*.

dragline   Equipment used to excavate and remove bottom materials from a water body. The materials are removed with a bucket that is pulled toward the piece of equipment with cables.

drain   (1) To remove and carry away surface or subsurface water. (2) An artificial channel or pipe used to transport surface or subsurface water. See *canal; ditch*.

drainage   (1) A watershed that contains all tributary rivers, streams, sloughs, ponds, and lakes that drain a given area. (2) The process of downward removal of surface and subsurface runoff water from soil either by gravity or artificial means.

drainage area   Total land area, measured in a horizontal plane, enclosed by a topographic divide, from which direct surface runoff from precipitation normally drains by gravity into a wetland, lake, or river. Also referred to as a catchment area, watershed, and basin. In the case of transbasin diversions, the drainage area would include water from all diverted streams. See *catchment area*.

drainage basin   The total surface land area drained by a stream or river from its headwater divides to its mouth. See *catchment basin*.

drainage basin density (D_d)   Ratio of total drainage channel length (miles or kilometers, L) to total drainage basin area (square miles or square kilometers, A):

$$D_d = L/A.$$

Synonymous with *stream density*.

drainage basin shape   See *drainage basin shape* under *dimensions*.

drainage divide   The boundary formed along a topographic ridge or along a subsurface formation that separates two adjacent drainage basins.

drainage lake   See *drainage lake* under *lake*.

drainage structure   A structure to remove runoff water composed of metal, concrete, or wooden culverts, open-faced culverts, bridges, and ditches.

drainage system   Natural or artificial channels that transport water out of a basin.

drainage texture   Expression of the space between stream channels above a reference point in a stream. Determined by dividing the number of stream crossings by the length of a contour.

drained   A condition where ground or surface water has been removed by artificial means to the point that an area no longer meets the hydrological criterion of a wetland.

draw   A long, wide topographic feature formed by perennial or intermittent surface runoff, without reference to the presence or absence of water.

drawdown   (1) Lowering of water levels stored behind a dam or other water control structure. (2) Change in reservoir elevation during a specified time interval. (3) Local decline of a water table due to water withdrawal.

dredge   (1) Act of sampling or excavating material from the bottom of a water body. (2) Equipment used for sampling or excavation material from the bottom of a water body.

drift   (1) Dislodgement of aquatic invertebrates and fish from a stream bottom into the water column where they move or float with the current. (2) Any detrital material transported by water current. (3) Materials deposited ashore by wind or water currents in a pond, lake, or reservoir. (4) Woody debris that has been modified by abrasion in a stream. Sometimes, this term refers to floating materials or surface water set in motion by the wind.

drift line   A visible elongated collection of floating debris, detritus, or organic matter that results when opposing forces such as wind and current meet along a bank contour (i.e., parallel to the shoreline of a lake or the flow of a stream) that marks the height of an inundation event or streamflow.

drift organism   Benthic organisms temporarily suspended in water and carried by the current of streams.

drop, drop structure   See *check dam* under *habitat enhancements*.

drop inlet   See *drop inlet* under **control structure**.

drop-off   A vertical or steep descent in the bottom of a water body.

drowned valley   Valley carved in land by a stream and later flooded by a rise in sea level.

drum gate   See *drum gate* under **control structure**.

drumlin   A tear-drop shaped landform that results from the deposit of glacial till or other drift with the tapered end pointing in the direction of the glacial movement.

dry dam   A dam designed for flood control that does not provide permanent water storage.

dry ravel   See *dry ravel* under **landslide**.

dry wash   An intermittent streambed in an arroyo or canyon that contains rainwater for a very short time.

Duboy's equation   An equation that calculates the force per unit of area exerted on the streambed at different flows or used to predict the capability of a system to move bed load material.

dug pond   See *dug pond* under **pond**.

dune   (1) Summit and sloping sides of a mound, hill, or ridge of loose, unconsolidated, granular windblown deposits. (2) In streams, bedforms that are generally transverse to the direction of flow with a triangular profile that advances downstream due to net deposition of particles from the gentle upstream slope to the steeper downstream slope. Dunes move down the stream at velocities that are small relative to streamflow.

dune bar   See *dune bar* under **bar**.

dune lake   See *dune lake* under **lake**.

dune pond   See *dune pond* under **pond**.

duration curve   A graphical representation (a curve) of the number of times given quantities such as streamflow occurred or a percentage of events within a time period. The event or quantity is arranged in order of magnitude along the ordinate with the time period expressed as a percentage along the abscissa.

duration of inundation or saturation   Length of time that water remains above the soil surface when water fills most of the interspaces between soil particles.

duty of water   In irrigation, the quantity of water required to satisfy the irrigation water requirements of land. It is expressed as the rate of flow required per unit area of land, the area which can be served by a unit rate of flow, or the total volumetric quantity of water in terms of depth, required during the entire or portion of the irrigation season. In stating the duty, the crop, the location of the land in question, and the soil type are usually specified.

dy   Soft, fine-grained sediment (composed almost entirely of organic matter) in lakes, bogs, marshes, or swamps. Yellow-brown, flocculent, fibrous, undecomposed plant material with a low pH that is derived to a great extent from peat in a bog sedge mat or from other allochthonous sources.

dynamic equilibrium   (1) Condition of a system where there is balanced inflow and outflow of material. (2) A migrating channel that is in equilibrium in terms of shape, pattern, and geometry.

dystrophic lake   See *dystrophic lake* under **lake**.

▶ e

early succession stage   See *early succession stage* under **succession**.

earth   (1) Soft surface materials composed of soil and weathered rock. See **ground, soil**. (2) Name for the Planet Earth.

earth dam   A barrier formed by the accumulation of earth that impedes the flow of water.

earthflow   See *earthflow* under **landslide**.

earth islands   See *earth islands* under **habitat enhancements**.

earthquake lake   See *earthquake lake* under **lake**.

earth slump   See *earth slump* under **landslide**.

ebullition   Bubbling up of gases produced through the decay of organic matter in the substrate.

ecoclimate Climate (temperature, humidity, precipitation, winds, and solar radiation) that occurs in a particular habitat or ecosystem.

ecocline Gradual, continuous change in an environmental condition of an ecosystem or a community.

ecosystem Any complex of living organisms interacting with nonliving chemical and physical components that form and function as a natural environmental unit.

ecosystem function Any performance attribute or rate function with some level of biological organization (e.g., energy flow, detritus processing, or nutrient spiralling).

ecotone Transition zone between two or more ecosystems or communities.

ectogenic meromixis See *ectogenic meromixis* under *mixing*.

edaphic Pertaining to or affected by geology or soil rather than climate or water.

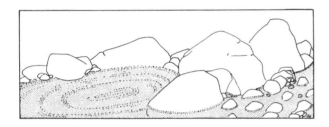

eddy (from Helm 1985)

eddy Circular current of water, sometimes quite strong, diverging from and initially flowing against the main current in streams. Eddies are usually formed where water flows past some obstruction or on the inside of river bends. Eddies often form backwater pools, alcove pools, or pocket water in rapids and cascades. See *eddy* under *slow water, pool, scour pool* under the main heading **channel unit**.

eddy flow Pattern of water movement created within an eddy.

eddy flux Continuous change and movement of water in an eddy.

eddy return channel See *eddy return channel* under *channel pattern*.

edge effect Ecotone or zone between two adjoin-

ing habitats or communities with different vegetation, climate, flow, or other features that contain an increased biological diversity and density of organisms.

edgewater Water that occurs at the interface between land and a water body. See also *edgewater* under *slow water* under the main heading **channel unit**.

effluent (1) Discharge of liquid into a water body or emission of a gas into the environment. Usually composed of waste material. For example, emission of combustion gases into the atmosphere from industry or manufacturing. (2) Also, may be used to describe a streamflowing out of a lake or reservoir.

effluent flow See *effluent flow* under *flow*.

effluent seepage Diffuse discharge of effluents into groundwater or surface water.

Ekman dredge Device for sampling macro-invertebrates in bottoms of water bodies with soft substrates.

Ekman spiral (1) Current drift that is deflected 45° from the direction of the wind by geostrophic deflection. (2) Sequence of direction shifts of current that is generated by wind, usually downward from a lake surface. Ekman spirals are progressively more clockwise in the Northern Hemisphere.

Ekman transport Refers to drifting currents or organisms in an Ekman spiral.

elevation Height above or below sea level at a given point on the earth's surface. Surface water level in lake or reservoir.

eluvial Of, related to, or composed of eluvium or fine weathered rock.

eluvium (1) Deposit of soil, dust, etc., formed by the decomposition of rock that is found at its place of origin. (2) Soil from which dissolved or suspended material has been removed by percolating water through the process of eluviation.

embankment An artificial deposit of natural material, such as a bank, mound, or dike, raised above the natural surface of the land, and erected along or across a wetted area to divert or hold back water, support a roadway, store water, or for other similar purposes. See **dike** and **streambank**.

embayment   (1) An area of water that is enclosed by the topography of the adjoining land. (2) Portion of a stream flooded seasonally or permanently by a reservoir. See **arm;** *backwater* under *slow water*, *dammed pool* under the main heading **channel unit; bay; cove.**

embeddedness   Degree that gravel and larger sizes of particles (boulders, cobble, or rubble) are surrounded or covered by fine sediment (e.g., less than 2 mm).

emergency spillway   Wide outlet in the dam of a reservoir that allows excess water to be passed or spilled during periods of high runoff. Often intended to protect the structural integrity of the dam when storage capacity is high.

emergent macrophyte   See *emergent macrophyte* under **macrophyte.**

emergent vegetation   Rooted aquatic plants with some herbaceous vegetative parts that project above the water surface. Also referred to as emersed vegetation.

emergent wetland   See *emergent wetland* under **wetlands.**

emulsion   Mechanical mixture of two liquids that do not readily mix, such as oil and water.

endemic   Species that is unique or confined to a specific locality or drainage.

end moraine   See *end moraine* under **moraine.**

endorheic   Interior drainage from which no or little surface water reaches the sea.

energy   Capacity for work or available power. In streams, the capacity of the water to mobilize and move materials.

  energy dissipation   Loss of kinetic energy in moving water due to bottom friction, pools, large rocks, debris, and similar obstacles that impede flow.

  potential energy (PE)   Amount of energy in precipitated water that will be dissipated during its transit to sea level:

$$PE = mgh \; ;$$

  m = mass of water;
  g = acceleration due to gravity;
  h = elevation above sea level.

shear stress   Ability of water to mobilize materials from the bed and banks in streams.

$$Y = \rho RS \; ;$$

  ρ = density of water;
  R = hydraulic radius;
  S = channel slope.

stream power (SP)   Stream power at a given location, expressed as joules per second per meter.

$$S\rho = \rho g Q S \; ;$$

  ρ = fluid density;
  g = acceleration due to gravity;
  Q = discharge;
  S = average gradient of the channel.

total stream power   Availability of energy at a given location along a stream in relation to temporal loss of potential energy:

$$P = \frac{m\Delta h}{\Delta t} \; ;$$

  m = mass of water;
  Δh = change in elevation above sea level;
  Δt = change in time.

unit stream power   Time-rate loss of potential energy per unit mass of water.

$$\omega = \rho g v s \; ;$$

  ρ = fluid density;
  g = acceleration due to gravity;
  s = energy slope of flowing water (usually assumed to be similar to channel gradient);
  v = velocity gradient.

energy dissipation   See *energy dissipation* under *energy*.

energy efficiency   Measure of work accomplished in comparison to energy expended.

enhancement   (1) An improvement of ecological conditions over existing conditions for aquatic, terrestrial, or recreational resources. (2) Any change that is made for the improvement of a structural or functional attribute for a species or habitat. Some enhancement activities that result in a positive impact on a single species or specific component of an ecosystem may negatively impact others.

enhancement flow   See *enhancement flow* under *flow*.

enrichment   Process where discharge or runoff carries nutrients into a water body, enhancing the growth potential for bacteria, algae, and aquatic plants.

entrainment   (1) Accumulation or drawing in of organisms by a current, such as at a power plant intake. (2) Progressive erosion of the lower layers in the thermocline that results in a corresponding lowering of the thermocline in a stratified lake.

entrenched channel   See *entrenched channel* under *confinement*.

entrenched stream   See *entrenched stream* under *stream*.

entrenchment   Stream channel incision from fluvial processes.

entrenchment ratio   Computed index used to describe the degree of vertical containment of a river. It is the channel width of a flood prone area at an elevation twice the maximum bank-full depth/bank-full width.

environment   Combination of physical, chemical, climatic, and biotic conditions that influence the development, growth, structure, and vigor of an organism, population, or community.

eolian   See *eolian* under *streambank material*.

epeirogenesis   Wide-ranging tectonic events that can raise large crustal blocks in the earth's surface and sometimes create large basins where lakes can form. Similar to orogeny (i.e., mountain building) but slower.

ephemeral   Short-lived or transitory.

ephemeral flooding   See *ephemeral flooding* under *flooding*.

ephemeral flow   See *ephemeral flow* under *flow*.

ephemeral lake   See *ephemeral lake* under *lake*.

epibenthos   Organisms living at the water–substrate interface.

epilimnion   See *epilimnion* under *stratification*.

epilithic   Refers to periphyton assemblages on stones. See *epilithic* under *periphyton*.

epilittoral   See *epilittoral* under *littoral*.

epineuston   (1) Microscopic organisms living on the upper surface of the air–water interface. (2) See *epineuston* under *neuston*.

epipelic   Occurring on the surface of fine sediment. See also *epipelic* under *periphyton*.

epiphyte   Plant that lives on another plant but is not parasitic.

epiphytic   See *epiphytic* under *periphyton*.

epipleuston   See *epipleuston* under *neuston* and *pleuston*.

epipotamon   See *epipotamon* under *potamon*.

epipsammic   See *epipsammic* under *periphyton*.

epirhithron   See *epirhithron* under *rhithron*.

epizooic   Refers to materials or organisms that live or dwell on the bodies of animals.

equilibrium drawdown   Ultimate, constant drawdown that results in a steady rate of pumped discharge.

equitability   Measure of the distribution of individuals among the species component in a diversity index. Also referred to as evenness.

erosion   (1) Process of weathering or wearing away of streambanks and adjacent land slopes by water, ice, wind, or other factors. (2) Removal of rock and soil from the land surface by a variety of processes including gravitation stress, mass wasting, or movement in a medium.

   accelerated erosion   Rate of erosion that is much more rapid than normal, natural or geologic erosion, due primarily to human activities.

   flow erosion   Rapid downhill movement of solum (i.e., soil above the A and B horizons of the parent soil), along with some parent soil material, in a highly saturated condition.

   fluvial erosion   Erosion caused by flowing water.

   geologic erosion   Normal or natural erosion from geologic processes through time.

   gully erosion   (1) Rapid erosion, usually in brief time periods, that creates a narrow channel which may exceed 30 m in depth. (2) Formation of gullies in surficial materials or bedrock by a variety of processes including erosion by running water, weathering and the impact of falling rock, debris slides, debris flows, and other types of mass movement; erosion by snow avalanches.

interill erosion   Uniform removal of soil from a small area by the impact of raindrops or by sheet flow.

natural erosion   Wearing away of the earth's surface by ice, water, wind, or other agents under natural environmental conditions.

normal erosion   Gradual erosion of land used by people that does not greatly accelerate natural erosion.

rill erosion   Erosion resulting from movement of soil by a network of small, shallow channels.

saltation erosion   Bouncing of small soil or mineral particles from the action of wind, water, or gravity.

sheet erosion   Erosion of soil from across a surface by nearly uniform action of rain or flowing water. Sometimes includes rill and interill erosion.

splash erosion   Dislodgement and transport of soil particles as a result of the impacts of raindrops.

surface creep erosion   Dislodgement and movement of small particles of soil or minerals as a result of wind or gravity.

surface erosion   Detachment and transport of surface soil particles by running water, waves, currents, moving ice, wind, or gravity. Surface erosion can occur as the loss of soil, in a uniform layer, in rills, or by dry ravel (broken or crumbled rock).

suspension erosion   Movement of dislodged soil or mineral particles by moving water or above the ground by air movement.

erosion cycle   Pattern of erosion in a given location that moves and deposits materials over time.

erosion feature   Physical feature or geological landmark created from differential erosion of adjoining materials.

erosion lake   See *erosion lake* under **lake**.

erosion remnant   Material remaining after erosion of the surrounding landform.

erosion surface   Land face that is being eroded.

escape ladder   See *escape ladder* under **habitat enhancements**.

escapement   Number of fish that survive natural and human-caused mortality to spawn.

escape ramp   See *escape ramp* under **habitat enhancements**.

escarpment   (1) Relatively continuous cliff or steep slope, produced by erosion or faulting, between two relatively level surfaces. Most commonly, a cliff produced by differential erosion. (2) A long, precipitous, cliff-like ridge of rock and other material, commonly formed by faulting or fracturing. See *scarp*.

essential elements   Collection of habitat elements, such as water, food, and shelter, that are essential to the continued existence of a plant or animal species; used to describe critical habitat for endangered species.

estuarine   See *estuarine* under **valley segments**.

estuarine zone   Environmental system consisting of an estuary and the transitional area that are consistently influenced or affected by water from sources such as, but not limited to, salt marshes, coastal and intertidal areas, bays, harbors, lagoons, inshore waters, and rivers.

estuary   That part of a river or stream or other body of water having unimpaired connection with the open sea, where the sea water is measurably diluted with freshwater derived from land drainage.

eulittoral   See *eulittoral* under **littoral**.

euphotic zone   Lighted region in a body of water that extends vertically from the surface to the depth at which light is insufficient to enable photosynthesis to exceed respiration of phytoplankton.

eupotamon   Lotic side arms of a river connected at both ends.

euryhaline   See *euryhaline* under **salinity**.

eurysaline   See *eurysaline* under **salinity**.

eutrophic   See *eutrophic* under ***trophic***.

eutrophication   (1) Natural process of maturing (aging) in a body of water. (2) Natural and human-influenced process of enrichment with nutrients, especially nitrogen (total nitrogen greater than 600 mg/m^3) and phosphorus (total phosphorus greater than 25 mg/m^3), leading to an increased production of organic matter. See also ***cultural eutrophication***.

eutrophy   State of the nutrient condition (especially nitrogen and phosphorus) in a water body. See ***trophic***.

evaporation   Loss of liquid water by transition to the gaseous phase.

evaporation pond   See *evaporation pond* under ***pond***.

evaporite   Remains of a solution after most of the solvent (usually water) has evaporated.

evapotranspiration   Movement of moisture from the earth to the atmosphere as water vapor by the evaporation of surface water and the transpiration of water from plants.

evorsion lake   See *evorsion lake* under ***lake***.

exorheic   An area of open basins whose rivers ultimately reach the sea.

extinction coefficient (η)   (1) Degree of light attenuation in water. (2) Availability of light with increasing depth as a negative exponential:

$$\eta = \frac{\log_e I_s - \log_e I_z}{z} \; ;$$

also given as:

$$\eta z = \log_e I_s - \log_e I_z \; ;$$

I_s = light at the water surface;
I_z = light at depth z;
z = depth.

extinct lake   See *extinct lake* under ***lake***.

▶f

fabric blanket   See *fabric blanket* under ***habitat enhancements***.

face   Slope that drains directly into the length of a stream without an obvious route of surface flow. These areas drain either by subsurface flow, shallow surface runoff without channels, or very small ephemeral streams.

faceted gully side   See *faceted gully side* under ***gully side form***.

facultative wetland plants   Plant species that usually occur in wetlands (estimated probability 67–99%), but occasionally found in uplands.

facultative wetland species   See *facultative wetland species* under ***wetland status***.

failing bank   See *failing bank* under ***bank stability***.

fall   A free, precipitous descent of water. The plural (falls) may apply to a single waterfall or to a series of waterfalls. See *falls* under *fast water—turbulent* under the main heading ***channel unit***.

fall overturn   See *fall overturn* under ***stratification***.

false color   See *false color* under ***remote sensing***.

false karst lake   See *false karst lake* under ***lake***.

fan   See ***alluvial fan, delta***.

fan apron   Accumulation of relatively recent alluvial material covering an older fan or piedmont.

fan terrace   An older, inactive alluvial fan that has partially eroded, with more recent accumulations of alluvial materials at the lower elevation.

farm pond   See *farm pond* under ***pond***.

fast water—nonturbulent   See *fast water —nonturbulent* under ***channel unit***.

fast water—turbulent   See *fast water—turbulent* under ***channel unit***.

fathom   A unit of measure equal to 1.83 m (6 ft).

fault   A plane or zone of fracture in a geographic feature (usually rock) that marks where permanent displacement or shifting has occurred.

fault sag pond   See *fault sag pond* under ***pond***.

feature   Any component of a community or ecosystem, obstruction, sample location, or other item plotted during aquatic biophysical mapping of streams and digitizing processes.

feeder   Term that is applied to a tributary stream.

fen   See *fen* under *wetlands*.

fence barrier   See *fence barrier* under *habitat enhancements*.

fetch   (1) An area where waves are generated by wind. (2) The distance waves travel in open water from their point of origin to the point of breaking. (3) The distance along open water or land over which wind blows without appreciable impedance or change in direction. (4) The distance across a lake from a given point to the upwind shore.

fill   (1) Localized deposition of material eroded and transported by a stream from other areas, resulting in a change in bed elevation. Compare with *scour*. (2) Deliberate placement of (generally) inorganic material on submerged land or low shoreline of a stream or other body of water to increase the surface elevation.

film water   Layer of water surrounding soil particles and varying in thickness from 1 to 100 or more molecular layers.

fine benthic organic matter   See *fine benthic organic matter* under *benthic organic matter*.

fine load   See *fine load* under *sediment load*.

fine particulate organic matter   See *fine particulate organic matter* under *organic particles*.

fines   Particulate material, less than 2 mm in diameter, including sand, silt, clay, and fine organic material.

fine sediment   Fine-grained particles (2 mm or less in diameter) in streambanks and substrate. *See substrate particle-size table*.

fine woody debris   Parts of woody vegetation, usually branches, twigs, leaves, roots, and smaller limbs.

finger lake   See *finger lake* under *lake*.

firth   Long, narrow indentations of a coastline.

fish attractors   See *fish attractors* under *habitat enhancements*.

fish depth   See *fish elevation*.

fish elevation   Elevation of a fish above a stream-bed measured at the tip of the fish's snout. See *focal point*.

fish habitat   Aquatic and riparian habitats that provide the necessary biological, chemical, and physical (i.e., environmental) requirements of fish species at various life stages.

fish habitat indices   See *habitat quality indices*.

fish ladder   Inclined waterway, commonly an artificial channel with stepped pools, installed at a dam or waterfall to allow passage of migratory fish over or around an obstruction. May also be referred to as a fishway, fish passageway, or fish pass.

fish sanctuary   Resting, rearing, or spawning area for fish where they are protected by barriers or regulation from exploitation by sport or commercial fishers.

fish screen   Screen placed at the entrance of a water diversion for irrigation or power generation to prevent entry of fish.

fish velocity   See *fish velocity* under *velocity*. Also referred to as focal point velocity.

fixed logs   See *fixed logs* under *large organic debris*.

fjord   Long, narrow inlet or arm along a sea coast bordered by steep cliffs or slopes that are usually formed through glacial erosion and typified by the presence of a partially obstructing rise in bathymetry at the seaward opening.

flap gate   See *flap gate* under *gate* and main heading *control structure*.

flashboard   Temporary barrier (usually constructed of wood), of relatively low height that is placed along the crest of a spillway on a dam to adjust the water surface in the reservoir. Flashboards are constructed so that they can be readily added to increase water storage during low water years or removed to decrease water storage during high water years.

flash flood   Rapid increase in streamflow due to surface runoff, usually caused by torrential upstream rainfall.

flash flow   See *flash flow* under *flow*.

flat bank   See *flat bank* under *streambank*.

flatland reservoir   Impoundment built in relatively level terrain or in a wide floodplain and characterized by large shallow areas, often where a meandering river channel remains with its associated oxbows and sloughs.

flat meander   See *flat meander* under *meander*.

flats   Level landform composed of unconsolidated sediments, usually mud or sand, that may be elongated or irregularly shaped and continuous with the shore, that is covered with shallow water or may be periodically exposed. May also be called a marsh or shoal. See *wetlands, shoal*.

floater   Tree or piece of timber that remains afloat in water.

floating aquatic vegetation   Rooted or free aquatic plants that wholly or partially float on the water surface.

floating macrophyte   See *floating macrophyte* under *macrophyte*.

floating meadow floodplain   See *floating meadow floodplain* under *floodplain*.

floats   See *floats* under *large organic debris*.

floc   Tuftlike mass of floating material formed by flocculation or the aggregation of a number of fine suspended particles.

flocculation   Coalescence of colloidal particles into a larger mass that precipitates in the water column.

flood   (1) Rising and overflowing of a water body onto normally dry land. (2) Any flow that exceeds the bank-full capacity of a stream or channel and flows onto the floodplain. (3) May also refer to an exceptionally high tide in marine waters. See *flooding*.

flood basin   Largest floodplain possible occurring within one drainage.

flood channels   Channels that contain water only during high water conditions or flood events. Also referred to as overflow channels.

flood control pool   Maximum storage capacity of a flood control reservoir.

flood control reservoir   Reservoirs built to store water during high water years to abate flood

damage. Extreme water level fluctuations usually characterize such reservoirs. See *reservoir*.

flood control storage   Water stored in reservoirs to abate flood damage. See *flood control pool*.

flood frequency   See *recurrence interval*.

flooding   Condition where the surface of soil that is generally dry is temporarily covered with flowing water from any source including inflow from streams during high water years, runoff from adjacent slopes, or any combination of sources. See *flood*.

   dam-break flooding   Downstream surge of water caused by the sudden breaching of a dam on an impoundment in a stream channel, that may be caused by a landslide, the deposit of a debris flow, or a debris jam.

   ephemeral flooding   Inundation of the floodplain that is short-lived or transitory.

   headwater flooding   Situation where an area becomes inundated directly by surface runoff from upland areas.

   long-duration flooding   Inundation from a single event ranges from seven days to one month.

   periodically flooded   Soils that are regularly or irregularly inundated from ponding of groundwater, precipitation, overland flow, stream flooding, or tidal influences that occur at intervals of hours, days, or longer.

   permanently flooded   Water regime where standing water covers a land surface during the entire year, except during extreme droughts.

flood level   Elevation or stage of water surface in a stream during a particular event when the stream exceeds bank-full capacity.

floodplain   (1) Area adjoining a water body that becomes inundated during periods of overbank flooding and that is given rigorous legal definition in regulatory programs. (2) Land beyond a stream channel that forms the perimeter for the maximum probability flood. (3) Strip of land bordering a stream that is formed by substrate deposition. (4) Deposit of alluvium that covers a valley flat from lateral erosion of meandering streams and rivers.

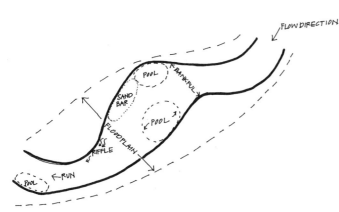

floodplain (from Firehock and Doherty 1995)

coastal delta floodplain Terminal lateral expansion of the alluvial plain where the river divides into numerous distributary channels as it enters an estuary or ocean.

floating meadow floodplain Flooded grassland forming vast floating mats of vegetation that may occur seasonally. May also apply to land areas that are semi-aquatic. See *wetlands*.

fringing floodplain Relatively narrow strip of floodplain between the walls of a river valley.

internal delta floodplain Floodplains created by geologic features that result from lateral spreading where the main channel divides and sheet flooding is common.

nonflooding floodplain Floodplain where overbank flooding occurs very rarely.

occasionally flooded Floodplain that is covered infrequently by standing water.

permanently flooded floodplain Floodplain that is permanently covered by water and that may contain floating, submerged, or rooted vegetation. See terms and definitions under *wetlands*.

seasonally flooded floodplain Floodplain that floods annually or seasonally.

floodplain processes See *floodplain processes* under active *valley wall processes*.

floodplain width Distance between the edge of the main stream channel and the land–water interface at maximum flood stage where flooding occurs an average of every two years.

flood pulse Periodic increase in riverine productivity that occurs when rivers inundate floodplains.

Nutrients, phytoplankton, and zooplankton produced in the floodplain provide a "flood pulse" when connectivity between the floodplain and river occurs.

flood pulse concept The pulse of river discharge due to flooding. See *flood pulse*.

flood recurrence interval See *recurrence interval*.

flood river See *flood river* under **river**.

floodway (1) Part of a floodplain that is contained by levees and is intended for emergency diversion of water during floods. (2) Part of a floodplain that is kept clear of obstructions to facilitate the passage of floodwater.

flood zone Land bordering a stream that is subject to floods of about equal frequency.

flotation load Total organic debris that enters water.

flotsam Floating debris of natural or human origin that is transported downstream but often collects in eddies.

flow (1) Movement of water and other mobile substances from one location to another. (2) Volume of water passing a given point per unit of time. Synonymous with *discharge*.

average annual discharge Product of the average water velocity of an aquatic system times the cross-sectional area of that system.

average annual inflow Sum of the mean annual discharge for all streams entering a water body.

base flow Portion of the stream discharge that is derived from natural storage (i.e., outflow from groundwater, large lakes or swamps), or sources other than rainfall that creates surface runoff; discharge sustained in a stream channel, not a result of direct runoff and without regulation, diversion, or other human effects. Also called sustaining, normal, dry weather, ordinary, or groundwater flow.

broken flow Nonlaminar flow with numerous standing waves.

channel maintenance or preservation flow Range of flows within a stream from normal peak runoff and may include, but is not limited to, flushing flows or flows required to maintain the existing natural stream channel and adjacent riparian vegetation.

creeping flow   Increased flow that spreads laterally across a low-relief, depositional floodplain.

duration flow   See *duration curve*.

effluent flow   Streamflow generated from groundwater.

enhancement flow   Increased streamflow from reservoir releases that improves natural streamflow conditions for aquatic, terrestrial, recreational, and other resources.

ephemeral flow   Streamflows in channels that are short-lived or transitory and occur from precipitation, snow melt, or short-term water releases.

flash flow   Sharp peaks in streamflows on a hydrograph.

flushing flow   Artificial or natural discharge of sufficient magnitude and duration to scour and remove fine sediments from the stream bottom that helps to maintain the integrity of substrate composition and the form of the natural channel.

generation flow   Flow that results from water releases for power generation.

improvement flow   Discharge that provides additional water for various uses, corrects the deterioration of water quality and utilization pressures, and results in increases in populations of aquatic organisms.

index flow   Standard measure of discharge used to compare other streamflows.

influent flow   Groundwater flows that recharge aquifers.

instantaneous flow   Discharge measured at any instant in time.

instream flow   Discharge regime for a stream channel.

instream flow requirements   Streamflow regime required to satisfy the water demand for instream uses.

intergravel flow   Portion of surface water that infiltrates a streambed and moves through the gravel substrate. Also referred to as interstitial flow.

intermittent flow   Flows that occur at certain times of the year only when groundwater levels are adequate but may cease entirely in low water years or be reduced to a series of separated pools. Compare with *ephemeral flow* under *flow*.

interstitial flow   See *intergravel flow* under *flow*.

laminar flow   Uniform streamflow with no mixing or turbulence.

least flow   Lowest flow established by agreement in a regulated stream that will sustain an aquatic population at agreed upon levels. This flow may vary seasonally. Compare with *minimum flow* under *flow*.

low flow   See *minimum flow* under *flow*.

maximum flow   See *peak flow* under *flow*.

mean annual flow   Average annual streamflow, usually expressed as cubic meters per second (m^3/s).

mean annual runoff   Mean annual flow divided by the catchment area (e.g, meters per hectare).

mean flow   Average discharge at a given stream location, usually expressed in cubic meters per second or cubic feet per second. The discharge is computed for the period of record by dividing the total volume of flow by the number of days, months, or years in the specified period. Also referred to as average discharge, mean discharge, average daily flow (ADF), or average annual flow (QAA).

minimum flow   (1) The lowest discharge recorded over a specified period of time. (2) Lowest flow established by agreement in a regulated stream that will sustain an aquatic population to agreed upon levels.

modified flow   Discharge at a given point in a stream resulting from the combined effects of all upstream and on-site operations, diversions, return flows, and consumptive uses.

natural flow   Stream discharge that occurs naturally through climate and geomorphology without regulation, diversion, or other modification by humans.

optimum flow   Discharge regime that provides the maximum flow for any specified use in a stream.

peak flow   Highest discharge recorded within a specified period of time that is often related to spring snowmelt, summer, fall, or winter flows. Also called maximum flow.

perennial flow   Flows that are continuous throughout the year.

placid flow   Flow that is slow, tranquil, and sluggish.

regulated flow   Streamflows that have been affected by regulated releases, diversions, or other anthropogenic perturbations.

return flow   Water, previously diverted from a stream, that is not consumed and returned to a stream or to another associated body of surface or groundwater.

rolling flow   Streamflows with numerous unbroken waves.

seasonal flow   Streamflows that exist only at certain times of the year, that may be derived from springs or surface sources, but usually are associated with seasonal precipitation patterns.

seven day low flow (Q7L)   Lowest discharge that occurs for a seven consecutive day period during the water year.

seven day/Q10 (Q7/10)   Lowest flow (i.e., a specific critical low flow) that occurs or is predicted to occur for seven consecutive days within a ten-year period.

sheet flow   Fast, nonturbulent flow over a level streambed that approaches laminar flow.

shooting flow   Swift, high energy streamflow with great erosive potential. Also referred to as supercritical flow. Compare with *tranquil flow*.

sluggish flow   Slow streamflow that occurs when runoff is spread over a period of time.

stable flow   Streamflow with a constant discharge.

steady flow   Little fluctuation in discharge during a specified period of time.

stem flow   Precipitation that is intercepted by plants and flows down a plant stem or tree trunk to the ground.

storm flow   Sudden and temporary increase in streamflow, resulting during a heavy rainfall that may be accompanied by additional water from increased snow melt.

streamflow   Flow of water, generally with its suspended load, in a well-defined channel or water course.

streamline flow   Tranquil flow slower than shooting flow. See *tranquil flow* under *flow*.

subsurface flow   Portion of streamflow moving horizontally through and below the streambed from groundwater seepage. It may or may not become part of the visible streamflow at some point downstream.

surface flow   Visible portion of streamflow that occurs above the substrate within a channel.

survival flow   Discharge required to prevent death of aquatic organisms in a stream during specified periods of time (e.g., seven days) when streamflows are extremely low.

swirling flow   Streamflows characterized by eddies, boils, and swirls.

tranquil flow   Slow streamflow with low energy that results in little erosion to streambanks and the stream bottom. Also referred to as *subcritical flow*. Compare with *shooting flow* under *flow*.

transition flow   Interface where laminar and turbulent flows meet.

transverse flow   Stream currents that operate at right angles to the main flow of the current. These currents are important in pool formation along the banks of meandering streams.

tumbling flow   Flow characterized by cascades, usually over large boulders or rocky outcrops.

turbulent flow   Streamflow characterized by an irregular, chaotic path, with violent mixing.

uniform flow   Flow where water velocities are the same in both magnitude and direction from point to point in a cross section. Uniform flow is possible only in a channel where the cross section and gradient are constant. Compare with *laminar flow* under *flow*.

unsteady flow   Flow in an open channel where the depth changes with time.

flow duration curve   Graphical representation of the number of times or frequency that a flow of a given magnitude occurs.

flow erosion   See *flow erosion* under *erosion*.

flow slides   See *flow slides* under *landslide*.

flow stability   A description of flow consistency at a given stream location.

flow till   See *flow till* under *landslide*.

flume   A chute constructed to transport materials with flowing water. See also *flume falls* under *fast water—turbulent, falls* under the main heading **channel unit**.

flush   (1) Rapid release of water from a storage structure, such as a reservoir, that has a sufficient quantity of water to transport suspended and floating materials. (2) Sudden rush of water down a stream that occurs during a freshet (flash flood). (3) Action of periodic high flow to keep a site at a spring or wetland wet or moist. (4) Rinsing or cleansing of beaches by a flush of water at high tide.

flushing flow   See *flushing flow* under *flow*.

flushing period   Period of time required for the total volume of water to be flushed through a system.

flushing rate   Time required for a volume of water equivalent to the reservoir volume to be discharged. The flushing rate is calculated by dividing the volume of the reservoir by the daily discharge.

fluvial   Pertaining to or living in streams or rivers, or produced by the action of flowing water. See also *fluvial* under **streambank material**.

   sediment production zone   Area of active erosion where sediments are derived and moved downstream. Also referred to as drainage basin.

   sink zone   Area with low gradient and water velocity where substrates are deposited. Also referred to as the zone of deposition.

   transportation zone   Area where the erosion and deposition are generally in equilibrium so that the total eroded material does not increase or decrease but is transported downstream. Also referred to as the transfer zone.

fluvial erosion   See *fluvial erosion* under *erosion*.

fluvial sedimentation   Sediment created, transported, and deposited by flowing water.

flux   Variation of substances, such as nutrients, per unit of time.

foam   Collection of minute bubbles on the surface of a liquid formed by agitation or fermentation.

focal point   Location, and the conditions at that location, occupied by an organism. Focal point measurements help to define microhabitat. See **microhabitat**.

fog   Accumulation of minute water particles in a layer at or just above the ground surface that often occurs in river bottoms, depressions, or low lying areas.

foliar cover   Percentage of ground or water covered by shade from the aerial portion of plants. Small openings in the canopy are excluded; foliar cover is always less than the canopy cover and is never greater than 100%. Synonymous with shading effect as measured with a vertical light source. See **canopy cover**.

foliar shading   See *foliar shading* under **stream surface shading**.

ford   Low-water stream crossing with bank access to allow wading or vehicular passage by people and crossing by livestock. The streambed must be composed of materials that are resistant to erosion. Streamflows must either be low enough to allow shallow crossings or the "ford" cannot be used during high-water periods.

forebay   (1) A reservoir or canal from which water is taken to operate equipment such as a waterwheel or turbine. (2) The entry chamber of a dam through which water flows into an outlet works. (3) The area of a reservoir immediately upstream and adjacent to the structure that forms a reservoir.

foreshore   (1) Land along the edge of a water body. (2) Portion of land between the high water mark and the low water mark.

forested wetland   See *forested wetland* under **wetlands**.

fork   The point at which a stream branches into two channels that may be of similar size and flow.

form   Shape of a streambank that results from fluvial processes.

formative discharge   See **dominant discharge**.

fossil water   Water accumulated in an aquifer during a previous geologic time that has an extremely long recharge period.

foul   Refers to materials or organisms that cling to a substrate so as to encumber it (i.e., to entangle or clog or render unsuitable for use).

fracture   Rock mass that is separated into distinct fragments or masses. In streams, fractures create discontinuous surfaces, often with distinct boundaries.

fragipan   Natural subsurface soil horizon with a high bulk density relative to the soil above that is brittle and appears cemented when dry but less so when moist. Generally such soils are mottled in color, only slowly permeable to water, and commonly exhibit bleached cracks that form polygons in horizontal section.

framework riffle   Riffle composed of coarse substrates, often resistant to hydraulic erosion, that does not accumulate much sediment.

frazil ice   Fine spicules of ice formed in water (i.e., slush) too turbulent to allow the formation of sheet ice or anchor ice.

Fredle index (f_i)   An index of the quality of salmonid spawning gravel, obtained by dividing geometric mean diameter of particle size by the sorting coefficient:

$$f_i = \frac{d_g}{S_o} \; ;$$

d_g = geometric mean particle diameter (see **geometric mean particle diameter**);
S_o = sorting coefficient, $(d_{75}/d_{25})^{\frac{1}{2}}$; d_{75} is the particle diameter at the 75th percentile of cumulative particle weight (weight is cumulated from small to large particles), and d_{25} is the diameter at the 25th weight percentile).

freeboard   (1) Vertical distance between the level of the surface of the liquid in a conduit, reservoir, tank, or canal, and the top of an open conduit, or levee that prevents waves or other movements by the liquid to overtop the confining structure. (2) Height of the sides in a boat above the water surface.

free-floating macrophyte   See *free-floating macro-phyte* under **macrophyte**.

free-flowing   Stream or stream reach that flows unconfined and naturally without impoundment, diversion, straightening, rip-rapping, or other modification of the waterway.

free-living   Plants or animals not attached to or parasitic on other plants or animals (i.e., capable of living and moving independently).

free log   See *free log* under **large organic debris**.

free meander   See *free meander* under **meander**.

free-swimming   Organisms that are actively moving or capable of moving in water.

free water   Water that drains freely by gravity.

french drain   Blind drain composed of piping with holes that allows water to seep into the piping and drain from the site.

frequency of inundation or saturation   Period that an area is covered by surface water or soil is saturated. It is usually expressed as the number of years the area is inundated or the soil is saturated during the portion of the growing season with prevalent vegetation.

frequently confined channel   See *frequently confined channel* under **confinement**.

frequently flooded channel   A channel that floods often (e.g., in most years).

freshet   Rapid temporary rise in stream discharge and level caused by heavy rains or rapid melting of snow and ice.

freshwater   See *freshwater* under **salinity**.

freshwater marsh   See *freshwater marsh* under **wetlands**.

fringe marsh   See *fringe marsh* under **wetlands**.

fringing floodplain   See *fringing floodplain* under **floodplain**.

Froude number (F_r)   Dimensionless number expressing the ratio of inertial to gravitational forces in a fluid:

$$F_r = \frac{V}{(gd)^{\frac{1}{2}}}$$

or

$$F_r = V^2/gD \; ;$$

V = mean velocity (m/s);
g = acceleration due to gravity (m/s^2);
D = hydraulic depth (m);

Values of F_r less than 1 are termed subcritical, and are characteristic of relatively deep, slow stream-flow. Values of 1 denote "critical flow." Values greater than 1 are termed supercritical, and are characteristic of shallow, fast water.

▶ g

gabion   See *gabion* under **habitat enhancements.**

gaging station   Particular location on a stream, canal, lake, or reservoir where systematic measurements of streamflow or quantity of water are made.

gaining stream   See *gaining stream* under **stream.**

gallery   See *gallery* under **control structure.**

gallery forest   Strip of forest confined to a stream margin or floodplain in an otherwise unforested landscape.

gate   See *gate* under **control structure.**

generation flow   See *generation flow* under *flow.*

geographic information systems   Systems for identifying locations geographically and organizing information about those locations in a relational process based on shared geographic location. Data are referenced with geographic coordinates and stored in digital format in a computer.

geologic erosion   See *geologic erosion* under **erosion.**

geometric mean diameter particle (d_g)   Measure of the mean particle size of substrate materials that is sometimes used as an index of spawning gravel quality; it is also referred to as the D_{50} size. It has been calculated in two ways:

$$d_g = d_1^{w1} \times d_2^{w2} \times \ldots d_n^{wn} \; ;$$

$d_1, \ldots d_n$ = particle diameters in size percentiles $1, \ldots, n$;
$w_1, \ldots w_n$ = decimal weight fractions in size percentiles $1, \ldots, n$.

Alternatively,

$$d_g = (d_{16} \times d_{84})^{1/2} \; ;$$

d_{16} = particle diameter at the 16th size percentile;
d_{84} = particle diameter at the 84th size percentile.

geometric registration   See *geometric registration* under **remote sensing.**

geomorphic   Pertains to the form or shape of the landscape and to those processes that affect the surface of the earth.

geomorphic surface   Unit on the earth's surface with a characteristic form that indicates the unit was formed by a particular process in a particular geomorphic environment.

geomorphological processes   Dynamic actions or events that occur at the earth's surface due to natural forces resulting from gravity, temperature changes, freezing and thawing, chemical reactions, seismic activity, and the forces of wind, moving water, ice, and snow. Where and when a force exceeds the strength of the earth material, the material is changed by deformation, translocation, or chemical reaction.

geomorphology   Study of the origin of landforms, the processes that form them, and their material composition.

georeferencing   Process of assigning map coordinates to image data.

geostrophic effect   Deflection (coriolis) force of the earth's rotation.

geothermal   Pertains to the internal heat of the earth. Frequently expressed as hot water springs, steam, and hot brines that seep to the surface or are forced into the air by hot groundwater generated by the internal heat of the earth. See *geyser.*

geyser   Hot water spring that intermittently erupts as a fountain-like jet of water and steam.

glacial drift   Rock debris transported and deposited by glacial ice or melt water.

glacial flour   Inorganic material pulverized to silt- and clay-size particles by the movement of glaciers and ice sheets. Term is also used to describe suspended or deposited glacial silt.

glacial lake   See *glacial lake* under *lake.*

glacier   Body of ice formed by the compaction and recrystalization of snow that has definite lateral limits and exhibits motion in a definite direction.

> **rock glacier**   Tongue of rock fragments held together by interstitial ice that moves downslope, similar to a glacier.

> **valley glacier**   Glacier that is smaller than a continental glacier or an ice cap that flows mainly along well-defined valleys, often with many tributaries.

glaciofluvial   Pertains to channelized flow of glacier meltwater as well as deposits and landforms formed by glacial streams.

glaciofluvial materials   Materials such as sediments deposited by glacial streams either directly in front of, or in contact with, glacial ice.

glaciolacustrine materials   Range of particles resulting from glacial action, deposited along the margins of glacial lakes or released into the lakes by melt water.

glaze   Homogeneous, transparent ice layers that are formed from rain or drizzle when the precipitation comes in contact with surfaces at temperatures of 0°C (32°F) or lower.

gleization   Process in saturated or nearly saturated soils that involves the reduction of iron, its segregation into mottles and concretions, or its removal by leaching. Gleizated soils tend to be grey in color when the iron has been reduced and reddish when the iron is oxidized.

glide   See *glide* under *slow water* under the main heading *channel unit.*

global positioning system (GPS)   System of satellites in permanent orbit above the earth that allow a receiver to triangulate their position on or above the earth's surface.

gorge   (1) Small, narrow canyon with steep, rocky walls, especially one through which a streamflows. (2) Portion of the water drainage system situated between the catchment area and the outwash fan.

graben   Portion of the earth's crust bounded on at least two sides by faults that have moved downward as a result of crustal activity to create an area that is lower than the adjoining landform.

graben lake   See *graben lake* under *lake.*

graded bed   Streambed with a series of sorted layers where the larger sizes of particles occur on the bottom and finer sizes on the surface.

graded stream   See *graded stream* under *stream.*

grade-stabilization or control structure   See *check dam* under *habitat enhancements.*

gradient   (1) General slope, or the change in vertical elevation per unit of horizontal distance, of the water surface in a flowing stream. (2) Rate of change of any characteristic per unit of length. See *profile.*

gravel   Substrate particle size between 2 and 64 mm (0.1 and 2.5 in) in diameter. Compare with other substrate sizes under *substrate size.*

gravel bed   (1) Natural accumulation or deposition of gravel-size particles in areas of low water velocity, decreased gradient, or channel obstruction. (2) Gravel artificially placed in a water body for fish habitat, primarily for spawning.

gravel pit lake   See *gravel pit lake* under *lake.*

gravel restoration   See *gravel restoration* under *habitat enhancements.*

gravity flow   See *gravity flow* under *landslide.*

gravity waves   See *gravity waves* under *wave.*

graywater   Drainage of dishwater, shower, laundry, bath, and washbasin effluents that contrast with "blackwater" such as toilet drainage.

greenway   Protected linear open-space area that is either landscaped or left in a natural condition. May follow a natural feature or landscape, such as river or stream, or other types of right-of-ways.

grit   Fine, abrasive particles, usually composed of coarse-grained siliceous rock (the size of sand or smaller) with a sharp, angular form and deposited as dust from the air or from impurities in water.

gross primary production   See *gross primary production* under *production.*

ground   Term applied to the solid surfaces of soil, weathered rock, and detritus on the earth. See *earth, soil*.

ground cover   See *ground cover* under **remote sensing**.

ground moraine   See *ground moraine* under **moraine**.

ground truthing   See *ground truthing* under **remote sensing**.

groundwater   (1) Water located interstitially in the substrate of the earth that is recharged by infiltration and enters streams through seepage and springs. (2) Subsurface water in a zone of saturation, standing in or passing though (groundwater flow) the soil and the underlying strata.

average linear velocity   The average linear velocity (V) of groundwater flow is estimated by:

$$V = \frac{K\,(d_h/d_l)}{n_e} \; ;$$

K = hydraulic conductivity (m/d);
d_h/d_l = hydraulic head;
n_e = effective porosity, the volume of interconnected pore space relative to the total volume.

coefficient of storage   The coefficient of storage, S, is estimated by:

$$S = \rho g h(\alpha + n\beta) \; ;$$

ρ = density of fluid;
g = gravitational constant;
h = saturated thickness of aquifer;
α = compressibility of the aquifer skeleton;
n = porosity;
β = compressibility of the fluid.

Darcy's law   Darcy's law describes the liquid transporting properties of porous materials:

$$Q = KA(d_h/d_l) \; ;$$

Q = flow rate (m^3/d);
K = hydraulic conductivity (m/d);
A = cross-sectional area through which flow occurs (m^2);
d_h/d_l = fluid head.

porosity   Porosity, n, is estimated by:

$$n = S_y + S_r \; ;$$

S_y = specific yield of groundwater;
S_r = specific retention (the volume of groundwater that does not drain under influence of gravity).

groundwater budget   Summation of water movement into and out of the groundwater during a specified period of time.

groundwater dam   Underground obstruction hindering the movement of groundwater.

groundwater discharge   Seepage or flow of water from groundwater to surface waters.

groundwater interchange   Pattern of recharge and discharge between groundwater and surface water.

groundwater level   Elevation of water in groundwater in relation to the surface or another fixed point.

groundwater recharge   Flow of water from the surface into groundwater.

groundwater runoff   Flow of groundwater along a geologic gradient by gravity that is comparable to surface runoff.

groundwater table   Surface below which rock, gravel, sand, or other substrates are saturated by a body of unconfined groundwater.

groyne (groin)   See groyne under **habitat enhancement**.

gulch   Small, narrow, often steep-sided stream valley that results from a secondary incision into a broader alluvial valley. See **ravine**.

gulf   Portion of an ocean or sea that is partly enclosed by land.

gully   Small valley, ravine, or an ephemeral stream channel that carries water during and immediately after rain that is generally longer than wide.

continuous gully   Gully with many finger-like extensions into its headwaters area that gains depth rapidly downstream and maintains the depth to the mouth of the gully.

discontinuous gully   Gully that begins as an abrupt headcut and may occur singly or as a chain of gullies following one another in a downslope direction.

gully erosion   See *gully erosion* under **erosion**.

gully plug   See *gully plug* and *check dam* under *habitat enhancements*.

gully side form   Shape or profile of gully sides compared to right angles in the direction of flow.

> **benched**   Sides having a horizontal section (less than 5% or at least 30 cm wide) in an otherwise vertical wall.

> **faceted**   Sides with various combinations of vertical and sloping segments.

> **sloping**   Sides with a general slope less than 65% from horizontal.

> **vertical**   Sides with the general slope greater than 65% from horizontal.

guzzler   Water entrapment and containment structure used primarily to provide water for wildlife and livestock in arid regions.

gyrals (gyres)   Surface currents in very large lakes or seas that circulate in very large swirls.

gyttja   Sediment mixture of particulate organic matter, often zooplankton fecal pellets, inorganic precipitates, and deposited matter that is less than 50% inorganic by dry weight. Consists of a mixture of humus material, fine plant fragments, algal remains, quartz grains, diatom fistulas, exoskeleton fragments, spores of pollen relics, and plankton.

▶ h

habitat   Specific type of place within an ecosystem occupied by an organism, population, or community that contains both living and nonliving components with specific biological, chemical, and physical characteristics including the basic life requirements of food, water, and cover or shelter.

habitat component   Single element (such as velocity, depth, or cover) of the habitat or area where an organism lives or occurs. Component is synonymous with attribute.

habitat diversity   Number of different types of habitat within a given area.

habitat enhancements   Actions taken to modify or enhance habitats to benefit one or more species.

> **a-jacks**   Structure with three arms fastened together, resembling a child's toy jack, made of concrete or metal and placed along a bank or shoreline to prevent erosion from waves, breakers, or current.

> **articulated concrete mattress**   Collection of concrete slabs wired together to form a large mattress that is used to stabilize a bank.

> **artificial channels**   Short channels designed for spawning or rearing fish that are located near but separate from stream channels.

> **artificial holes**   Cavities of tile, pipe, hollow logs, and similar structures of appropriate size that are plugged at one end and are for fish spawning.

> **artifical meander**   A human-constructed stream channel that resembles a natural meander.

> **artificial reef**   Structure of concrete, tires, or other solid material constructed to create cover for aquatic organisms in marine or large freshwater environments.

> **artificial riffle**   Stream reach where rocks are added to create a shallow area with turbulent flow.

> **baffle**   Any device that changes the direction and distribution of flow or velocity of water.

> **bank stabilization**   Placement of materials such as riprap, logs, gabions, and planting of vegetation to prevent bank erosion.

> **berm**   Natural or artificial levee, dike, shelf, ledge, groyne, or bench along a streambank that may extend laterally along the channel or parallel to the flow to contain the flow within the streambank. Also referred to as *bund*. See *dike, levee*.

> **brushpile**   Trees, brush, or other vegetation tied into bundles and placed in a water body as supplemental structure or cover.

> **bulkhead**   Wall of wood, metal, rock, or concrete used to support a slumping bank.

> **channel constrictor**   Type of structure, such as a notched drop structure or a double-wing

current deflector, that forces the main flow of water through a narrow gap.

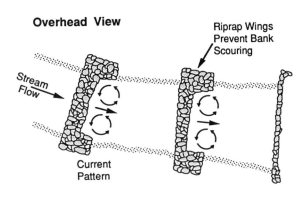

Overhead View

Riprap Wings Prevent Bank Scouring

Stream Flow

Current Pattern

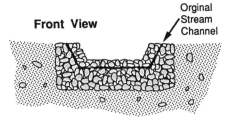

Front View

Orginal Stream Channel

Longitudinal Cross Section

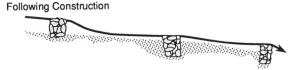

Following Construction

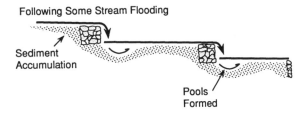

Following Some Stream Flooding

Sediment Accumulation

Pools Formed

check dams (from Meehan 1991)

check dam   Small dam of logs, rock, gabions, concrete, or other materials that completely spans the stream channel, slowing swift current and creating a plunge pool downstream from the sudden drop in channel elevation. Used to control soil erosion, headcutting in streams, and retard the flow of water and sediment in a channel.

cover   Logs, boulders, brush, or other materials placed along the banks, on the bottom, or

midwater, in a water body to create habitat for spawning or nursery areas.

crib   Cubical structure constructed of logs, metal posts, boulders, or other materials that is filled with rocks to protect streambanks and enhance cover in streams, lakes, or reservoirs. See *gabion* under **habitat enhancements**.

cribwell   Log crib that is anchored to the bank that is usually filled with rocks or dirt and planted with woody vegetation such as willows.

deflector   Single or double wing dam or jetty constructed from boulders, logs, or gabions that is used to force the current in a different direction. See *gabion* and *jetty* under **habitat enhancements**.

digger log   Type of current deflector that consists of a log anchored to a streambank that protrudes into or across the channel to deflect the flow and excavate a depression in the substrate.

drop, drop structure   See *check dam* under **habitat enhancements**.

earth islands   Islands of earth constructed in lakes, ponds, reservoirs, or other water bodies that provide habitat for birds or terrestrial animals and shallow-dwelling aquatic organisms.

escape ladder   Ladder-like structure of wood or mesh that provides animal access to developed water sources.

escape ramp   Structures (ramps) of rock, concrete, or wood that are placed along steep sides of ponds, canals, or conduits to permit animals access to water or to escape if trapped.

fabric blanket   Flexible, mesh-like material (usually synthetic), that reduces erosion and through which vegetation can grow. Sometimes placed under channel structures or riprap to hold materials.

fence barrier   Fence-like arrangement of logs, pilings, gabions, or other materials placed along a streambank.

fish attractors   Brush, tires, plastic, stake beds, clay pipes, standing timber, or other structures used to create habitat for fish or other organisms.

gabion   Wire cage or basket filled with rocks or

stone used to stabilize banks and to enhance aquatic habitat.

grade-stabilization or control structure   See *check dam* under **habitat enhancements**.

gravel restoration   Placement of gravel in a stream or other water body as spawning habitat for fish.

groyne (groin)   (1) Structure of wood, stone, or concrete projecting into water to prevent sand, pebbles, gravel, or other substrates or materials from being washed away by the current. (2) Structure usually constructed of logs or rocks and placed on a streambank to control the soil erosion or direction of water current.

gully plug   (1) Refers to the construction of check dams in gullies. See *check dam* under **habitat enhancements**. (2) Also refers to filling of gullies with soil and planting vegetation.

half-log   Log split lengthwise and anchored to the substrate (split side down) so that there is a gap between the log and the substrate that serves as cover for fish.

Hewitt ramp   Type of check dam made of logs, wood planks, or rocks that creates a gradual incline downstream so that water can excavate a plunge pool in a stream with little gradient. See *check dam* under **habitat enhancements**.

hurdles   Smaller branches woven together and pegged against a bank on a moderate slope to control bank erosion and promote growth of vegetation.

jack   See *a-jacks* under **habitat enhancements**.

jetty   Structure of rock, logs, pilings, gabions, or other materials that projects part way across a channel to direct the stream current.

k-dam   Log structure with a "K" shape constructed in a stream channel as habitat for fish and other aquatic organisms.

leg-type structure   Concrete structure, supported by either two legs along one side or four legs at the corners, or a "+" in the middle, placed in a stream channel.

log sill   See *check dam* under **habitat enhancements**.

low-head dam   See *check dam* under **habitat enhancements**.

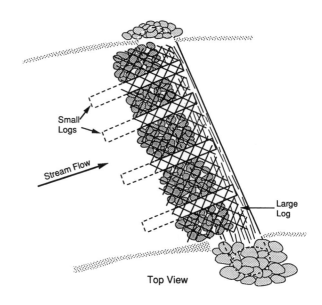

Top View

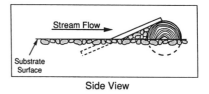

Side View

log sill (from Meehan 1991)

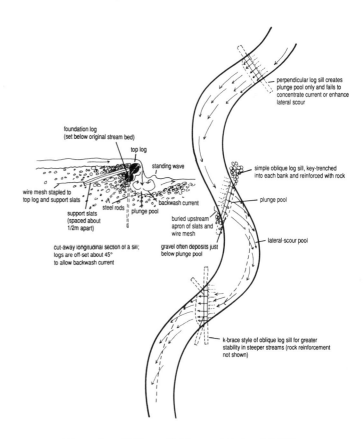

oblique log sill (from Kohler and Hubert 1993)

lunker structure   Plank and log, free-standing, box-like structure with open sides that is installed just below the water at the toe of a bank to stabilize the bank and create habitat.

mudsill   Logs placed perpendicular to the bank of a stream to prevent erosion.

off-channel pools   Pools or ponds created in the riparian area adjoining a stream channel and connected with the channel, creating a protected area for rearing fish or other aquatic organisms.

retard   See *fence barrier* under *habitat enhancements*.

revetment   A facing or structure made of large, durable material, such as earth-filled sacks, trees, logs, stumps, gabions, or rocks placed on a streambank to deflect current. Term also refers to the materials used to construct the revetment. Similar to *riprap* under *habitat enhancements*.

riprap   Hard materials, such as logs, rock, or boulders (often fastened together) used to protect a bank or another important feature of a stream, lake, reservoir, or other water body. Similar to *revetment* under *habitat enhancements*.

roller dam   See *check dam* under *habitat enhancements*.

scour structure   Structure of rocks, logs, gabions, or other materials placed along a streambank or in a stream channel to cause scouring and deposition of substrates.

sediment trap   Artificial pool that is generally excavated or constructed for collecting or trapping sand, silt, or other substrates that are transported as the streambed load.

sill   Low horizontal barrier of rock or other durable material in a stream. More particularly, a low barrier that is generally constructed of masonry across a gorge to control streambed erosion. Compare with *gully plug* and *check dam* under *habitat enhancements*.

spawning box   Boxes filled with gravel that are placed in shallow water of a lake or reservoir as artificial spawning habitat for fish.

spawning marsh   Artificial marshes constructed along a stream channel or margin of a lake or reservoir as spawning habitat for fish.

spawning platform   A floating or suspended wooden platform placed in a lake or reservoir as spawning habitat for fish.

spawning reef   An artificial reef of gravel, rock, or other material to serve as a spawning area for fish.

spiling   Term applied to control soil erosion using stakes that are driven into the ground with branches woven between the stakes, or two rows of staggered posts with brush between them to help reduce soil erosion and promote growth of vegetation.

stake bed   Artificial fish habitat created either by driving narrow wooden stakes vertically into a lake bottom before impoundment (or when the water level is low) or by attaching similar wooden stakes to a weighted frame that is lowered into place after impoundment.

structure   Durable materials, such as large organic debris, boulders, rock, or concrete placed in a stream, lake, reservoir, or other water body to create habitat.

tetrapod   Structure made from four legs of pre-cast concrete joined at a central point, all at angles of 109.5° to each other, that functions similar to an a-jack. See *a-jack* under *habitat enhancements*.

timber crib   Crib constructed of timber or planks in or along the margin of a water body to create habitat for aquatic organisms or to stabilize a bank.

tire reef   Artificial structure made of discarded vehicle tires that are generally attached together and weighted to keep the reef on the bottom of a water body. Used to create fish habitat in fresh or saltwater. A type of *artificial reef* described under *habitat enhancements*.

training wall   A low dam, that is normally constructed parallel or at an angle to a streambank, to divert flow in a specific direction.

trash collector   Fence-like structure of heavy wire fastened to metal stakes or logs and placed across a stream to intercept and hold debris flowing downstream that creates a dam and a plunge pool. Used to protect bridge crossings, create pools as habitat for fish, and collect gravel for spawning habitat. Synonymous with debris catcher, grizzly, or trash catcher.

tree retards   Trees placed along a streambank to reduce current and accumulate debris.

wedge dam   Dam or weir constructed with the apex facing upstream.

weir   See *check dam* under *habitat enhancements*.

wing dam   See *jetty* under *habitat enhancements*.

Wisconsin bank cover   Artificial ledge of wood that is supported by rock on the margin of a stream channel as cover for fish.

habitat fragmentation   Division of existing habitats into separate discrete units from modification or conversion of the habitat by anthropogenic or natural activities.

habitat quality indices   Numerical indices have been devised to give an overall rating of the quality or quantity of an aquatic or terrestrial habitat. In aquatic habitats, these indices often integrate many habitat elements such as streamflow regime, substrate, cover, and water quality and are frequently used in stream condition assessments or evaluations and to predict environmental impacts.

habitat type   Aggregation of land or aquatic units having equivalent structure, function, and responses to disturbance and capable of maintaining similar animal or plant communities.

half-log   See *half-log* under h*abitat enhancements*.

haline   See *haline* under *salinity*.

haline marsh   See *haline marsh* under *wetlands*.

halocline   See *halocline* under *salinity*.

hanging garden   Community of plants and animals that develops on steep slopes or cliffs where groundwater seeps onto the surface to form a confined ecosystem.

hanging valley   (1) Valley formed by a tributary stream that has a noticeably higher elevation than the main channel at their confluence. (2) Valley that is high above a shore to a water body and was formed from rapid erosion of the land or displacement by a fault.

hardness   Total concentration of calcium and magnesium ions expressed as milligrams per liter

(mg/L) of calcium carbonate. Synonymous with total hardness.

hardpan   A layer of earth that has become relatively hard and impermeable, usually through mineral deposits. A chemically hardened layer where the soil particles are cemented together with organic matter of SiO_2, sesquioxides, or $CaCO_3$.

hardwater   Water with a large concentration of alkaline cations derived from carbonates, chlorides, and sulfates of calcium and magnesium that is generally found in areas of limestone substrate. Compare with *softwater*.

head   Energy per unit weight of a fluid such as that resulting from a difference in depth at two points in a body of water. Includes pressure head, velocity head, and elevation head. See *hydraulic head*.

head cut   Upstream migration or deepening of a stream channel that results from cutting (i.e., erosion) of the streambank by high water velocities.

headland   Point of land, usually high and with a sheer drop, extending out into a water body, especially a sea.

head race   Channel that delivers water to a water wheel and causes the wheel to turn.

headscarp   See *headscarp* under *landslide*.

headwall   Steep slopes at the upper end of a valley, water course, or the cliff faces around a cirque.

headwater flooding   See *headwater flooding* under *flooding*.

headwaters   (1) Upper reaches of tributaries in a drainage basin. (2) The point on a nontidal stream above which the average annual flow is less than five cubic feet per second. For streams that remain dry for long periods of the year, the headwaters may be established as the point on the stream where a flow of five cubic feet per second is equaled or exceeded 50% of the time.

headwater stream   See *headwater stream* under *stream*.

heat budget   (1) Amount of heat per unit of surface area necessary to raise a body of water from the minimum temperature of winter to the maximum temperature of summer. (2) Total amount of heat between the lowest and highest heat content.

summer heat income Amount of heat needed to raise a body of water from isothermal winter conditions of 4°C to maximum summer levels.

winter heat income Amount of heat needed to raise a body of water from its minimum levels to isothermal conditions at 4°C.

height See *height* under **wave**.

Henry's law At a constant temperature, the amount of gas absorbed by a given volume of liquid is proportional to the atmospheric pressure of the gas.

heterotrophic See *heterotrophic* under **trophic**.

Hewitt ramp See *Hewitt ramp* under **habitat enhancements**.

high gradient riffle See *high gradient riffle* under *fast water—turbulent, riffles* under the main heading **channel unit**.

highland reservoir Impoundment built on steep terrain with a narrow floodplain. Most of the shoreline is characterized by steep, rocky, or clifflike formations.

high moor See *high moor* under **wetlands**.

high tube overflow See *high tube overflow* under **control structure**.

holding pond See *holding pond* under **pond**.

hole Part of a water body or channel unit that is distinctly deeper than the surrounding area.

hollow Valley or landscape depression.

holomictic See *holomictic* under **mixing**.

homogeneous Refers to a water body that has a uniform chemical composition throughout.

homoiohaline See *homoiohaline* under **salinity**.

homothermous Refers to a water body that has the same temperature throughout.

hot brine Saline water exceeding 37°C (98°F).

hot spring See *hot spring* under **spring**.

hummock Rounded, undefined or chaotic pattern of steep-sided low hills and hollows.

hummocky moraine See *hummocky moraine* under **moraine**.

humus General term for the dark organic material of soils that is produced by the decomposition of vegetable or animal matter. The more or less stable fraction of soil from decomposed organic material, generally amorphous, colloidal, and dark-colored. Nonhumus organic material is generally of low molecular weight, easily degraded by microbes, and has a short retention time.

hurdles See *hurdles* under **habitat enhancements**.

hydraulic control point Top of an obstruction in a stream or a point in a stream where streamflow is constricted by any large, relatively immobile object (e.g., boulder, bedrock) that stabilizes the stream geometry and maintains long-term channel character.

hydraulic depth See *hydraulic depth (D)* under **dimensions**.

hydraulic discharge See *hydraulic discharge* under **discharge**.

hydraulic drop Point where the streamflow changes from subcritical to supercritical in a reach with rapid changes in flow.

hydraulic flushing rate Rate at which an entire volume of lake water is replaced. See **flushing rate**.

hydraulic gradient (1) Slope of water in a stream. (2) Drop in pressure head per unit length in the direction of the streamflow.

hydraulic head Height of the free surface in a water body above a given point beneath the surface, or the height of the water level at an upstream point compared to the height of the water surface at a given downstream location.

hydraulic jump Transition form of supercritical flow that usually results from a change in channel elevation, geometry, or gradient. It may be confined to a very short reach where the depth increases dramatically in a turbulent wavelike feature or in a more gradual or undulating increase.

hydraulic radius   Ratio of the area of flow divided by the wetted perimeter of a cross section:

$$R = A/P \, ;$$

A = area;
P = wetted perimeter of a cross section.

hydraulics   Refers to the motion and action of water and other liquids.

hydraulic slope   Change in elevation of water surface between two cross sections divided by the distance between them. Compare with *gradient*.

hydric   Moist.

hydric soils   Soils that are saturated, flooded, or ponded long enough during the growing season to develop anaerobic conditions in the upper layers.

hydrograph   Graph that illustrates the relation of discharge, stage velocity, or other water component with time, for a given point on a stream.

hydrographic maps   Maps of lakes with contour lines delineating depth.

hydrohaline   See *hydrohaline* under *salinity*.

hydrologic balance   Relationship between the quality and quantity of water inflow to, water outflow from, and water storage in hydrologic units such as an aquifer, drainage basin, lake, reservoir, or soil zone. It encompasses the dynamic relationships of precipitation, runoff, evaporation, and changes in surface and groundwater storage.

hydrologic budget   Compilation of the total water input to and output from a lake, drainage, or watershed.

hydrologic control   Natural or artificial structures or conditions that manage or control the movement of surface or subsurface flow.

hydrologic cycle   Cycle of water movement from the atmosphere to the earth via precipitation through surface and groundwater to the oceans and return to the atmosphere via evaporation from water bodies and transpiration by plants. See *water cycle*.

hydrologic feature   Refers to water-related features visible at the land surface, such as stream channels, seepage zones, springs, and soil moisture including soil moisture characteristics as deduced from vegetation characteristics.

hydrologic regime   Water movement in a given area that is a function of the input from precipitation, surface, and groundwater and the output from evaporation into the atmosphere or transpiration from plants.

hydrologic zone   Area that is inundated by water or has saturated soils within a specified range of frequency and the duration of inundation or soil saturation.

hydrometabolism   Metabolism that occurs in free waters that is essentially all aerobic.

hydromorphic soil   Soils that develop in the presence of water sufficient to create anaerobic conditions in the soil.

hydrophobic soils   Soils that can repel water due to the presence of materials, usually of plant or animal origin, or condensed hydrophobic substances.

hydrophyte   (1) Plants that grow in water or saturated soils. (2) Any macrophyte that grows in wetlands or aquatic habitats on a substrate that is at least periodically deficient in oxygen because of excessive water content.

hydropower reservoir   Artificial water storage impoundments designed to retain water and redirect it through turbines to generate electricity.

hydrosaline   See *hydrosaline* under *salinity*.

hydrothermal   Refers to water on or within the earth that is heated from within the earth.

hydrothermal alteration   Process where heated groundwater passes through cracks or pores in rock that may alter the water, mineral, and rock.

hydrothermal vein   A cluster of minerals in a rock cavity that were precipitated by hydrothermal activity.

hygric   See *hydric*.

hygrophyte   Plants restricted to growing in or on moist sites. See *hydrophyte*.

hygroscopic water   Water adsorbed by dry soil from an atmosphere of high relative humidity.

hypereutrophic   See *hypereutrophic* under **trophic**.

hypertrophic   See *hypertrophic* under **trophic**.

hypoeutrophic   See *hypoeutrophic* under **trophic**.

hypolimnion   See *hypolimnion* under **stratification**.

hypopleuston   See *hypopleuston* under **pleuston**.

hypopotamon   See *hypopotamon* under **potamon**.

hyporheic zone   Latticework of underground habitats through the alluvium of the channel and floodplain associated with streamflows that extend as deep as the interstitial water in the substrate.

hyporhithron   See *hyporhithron* under **rhithron**.

hypothermal   Refers to water on or within the earth that is lukewarm or tepid.

hypotrophic   See *hypotrophic* under **trophic**.

hypsographic curve   Depth–area curve that describes the relationship of the cross-sectional area of a lake at a specific depth. A depth–volume curve is closely related and describes the volume of the lake at a specific depth.

▶ i
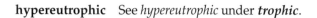

ice   See *ice* under **streambank material**.

iceberg   Large floating mass of ice.

icecap   Ice covering an area that slopes in all directions from the center unless confined between steep slopes.

ice dam   Dam created in the spring when broken sections of ice create a structure that blocks or retards the movement of water.

ice-disintegration moraine   See *ice-disintegration moraine* under **moraine**.

ice jam   Large blocks and pieces of ice that are jammed by a channel constriction or obstruction, usually occurring during spring breakup. If water cannot penetrate freely through the ice, a dam can form that may cause flooding, damage to human-built structures, or loss of human lives. See *ice dam.*

ice rafting   Sections of ice that form and float downstream when ice covering a river breaks up.

ice scouring   Abrasion or erosion of stream bottoms by ice that normally occurs from ice jams.

ice wedge   Accumulation or piling of ice that is pushed forward by the flow of water or pressure from ice and gravity.

IFIM   Abbreviation for instream flow incremental methodology. A method for relating changes in the physical characteristics of a stream to changes in flows.

illumination   Amount of light radiation impinging on a surface, expressed as lux, watts per square centimeter, foot-candles or lumens per square meter.

illuviation   Transport and deposition of soil components from a higher to a lower level by percolating water.

image   See *image* under **remote sensing**.

imbrication   Shingled or downstream overlapping of bed material due to water flow, most commonly as plate-shaped large gravel or cobble substrate materials.

impermeable   Refers to a layer of material of sufficient composition, density, and thickness that it does not permit passage of a liquid or a gas.

impervious   Refers to material through which water cannot pass or passes with great difficulty.

impoundment   Natural or artificial body of water that is confined by a structure such as a dam to retain water, sediment, or wastes.

improvement flow   See *improvement flow* under **flow**.

impurity   A foreign, objectionable material that contaminates water.

incidental drift   Casual or random drift of aquatic organisms.

incident light   See *incident light* under **solar radiation**.

incised channel   See *incised channel* under **channel geometry**.

incised stream   See *incised stream* under **stream**.

index   Measurement of feature of an organism, community, or habitat that is used as a reference for determining or monitoring change over time.

index flow   Standard discharge that is used to compare with other discharges in a specific stream. See also *index flow* under **flow**.

index of biotic integrity (IBI)   See index of biotic integrity (IBI) under **biological indices**.

index of refraction   (1) Number indicating the speed of light in a given medium versus the speed in water or other specified medium. (2) Measure of the degree to which light is bent or scattered when passing through water or other substances.

indicator organisms   (1) Organisms that respond predictably to various environmental changes, and whose presence, absence, and abundance are used as indicators of environmental conditions. (2) Any plant or animal that, by its presence, its frequency, or its vigor, indicates any particular property of a site.

indigenous   A fish or other aquatic organism native to a particular water body, basin, or region.

indirect toxicity   Toxicity that affects organisms by interfering with their food supply or modifying their habitat instead of acting directly on the organisms.

infiltration   Process by which water moves from the earth or surface water into the groundwater system.

infiltration rate   Rate at which standing water percolates downward into the substrate.

inflow   Location where water from one source enters another water body. Also, the movement of water from one source into another water body.

influent flow   See *influent flow* under **flow**.

influent seepage   Water movement from the surface of the ground toward the water table. Also, refers to seepage of water into the streambed from a stream above the water table.

initial breaching   See *initial breaching* under **breaching**.

initial dilution zone   Area receiving water adjacent to a point source discharge, extending from the point of discharge 100 m in all directions from the water surface to the bottom. State and federal water quality objectives do not apply within an initial dilution zone.

inlet   Long and narrow indentation of a shoreline or a narrow passage between two islands.

inner gorge   Stream reach bounded by steep valley walls. Common in areas of stream downcutting or geologic uplift.

inshore   On or very close to the shore. Also referred to as onshore.

insolation   Exposure to sunlight. Solar radiation is measured as the rate of delivery or intensity per unit area on the horizontal surface of an object.

instanteous discharge   See *instanteous discharge* under **discharge**.

instanteous flow   See *instanteous flow* under **flow**.

instream   Within the wetted perimeter of the stream channel.

instream cover   Areas with structure (e.g., boulders, rocks, logs, etc.) in a stream channel that provide aquatic organisms with shelter or protection from predators or competitors. Also a place with low water velocity where organisms can rest and conserve energy.

instream flow   See *instream flow* under **flow**.

instream flow requirements   See *instream flow requirements* under **flow**.

insular   Of or pertaining to an island or islands. Also applied to dwelling on an island, forming an island, or occurring alone.

insulated stream   See *insulated stream* under **stream**.

intensity   See *irradiance*.

interbasin transfer   Transport of water from one watershed or river basin to another.

interception   Physical interference of precipitation by vegetation.

intercratonic basin Basin formed in a relatively rigid and immobile area of the earth's crust.

interface Surface that forms a boundary between adjacent areas, bodies, or spaces. May also refer to the point of interaction between two independent systems.

interflow (1) Precipitation that infiltrates the soil and moves laterally until it resurfaces or is intercepted by an underground body of water. (2) Movement or inflow of water in a reservoir or lake between layers of water of different (higher or lower) density. See also *interflow* under **mixing.**

interfluve Elevated region that divides watersheds.

intergravel flow See *intergravel flow* under **flow.**

interill erosion See *interill erosion* under **erosion.**

intermediate breaching See *intermediate breaching* under **breaching.**

intermittent (1) Alternately starting and stopping. (2) Water that flows or exists sporadically or periodically.

intermittent flow See *intermittent flow* under **flow.**

intermittent lake See *intermittent lake* under **lake.**

intermittently exposed See *intermittently exposed* under **water regime.**

intermittently flooded See *intermittently flooded* under **water regime.**

internal delta floodplain See *internal delta floodplain* under **floodplain.**

internal progressive wave See *internal progressive wave* under **wave.**

internal seiche See *internal seiche* under **wave.**

interrupted stream See *interrupted stream* under **stream.**

interstitial flow See *intergravel flow* under **flow.**

interstitial space Spaces or openings in substrates that provide habitat and cover for benthos.

interstitial velocity Rate of subsurface water flow through the substrate (V_a), expressed as the volume of flow per unit of time through a unit area composed of solids and voids. Also referred to as apparent or intragravel velocity:

$$V_a = KS \; ;$$

K = permeability (cm/h);
S = hydraulic gradient.

It can also be approximated from surface velocities by the equation:

$$V_a = \frac{(V_s)^2 n^2 K}{2.22 R^{4/3}} \; ;$$

V_s = mean velocity of surface water;
n = roughness coefficient;
K = permeability;
R = hydraulic radius.

intertidal Zone between high and low tides.

intrusive body Igneous rocks which, while fluid, penetrated into or between other rocks but solidified before reaching the surface; includes very hard, coarse-grained rocks like granite, diorite, and gabbro.

inundation To be covered with standing or moving water.

inverse estuary Type of estuary where evaporation exceeds the freshwater inflow plus precipitation.

inverse stratification See *inverse stratification* under **stratification.**

invert (1) Refers to the upstream end of a culvert. (2) The bottom inside surface of a pipe or conduit. Occasionally the term refers to the bottom or base elevation of a structure.

iones Relic riverbeds.

irradiance Amount of electromagnetic energy received at a unit surface area per unit time, usually measured for the visible and near-visible portion of the light spectrum. Light passing through a water surface is progressively absorbed with increasing depth; light intensity at depth z (I_z) is:

$$I_z = I_o e^{-\eta z}$$

I_o = light intensity at the water surface;
η = extinction coefficient of light in water.

irregular acute meander   See *irregular acute meander* under *meander*.

irregular channel   See *irregular channel* under *channel pattern*.

irregular meander   See *irregular meander* under *meander*.

irregular meander channel   See *irregular meander channel* under *channel pattern*.

irregular meander with oxbow   See *irregular meander with oxbow* under *meander*.

irrigation   Movement of water through ditches, canals, pipes, sprinklers, or other devices from the surface or groundwater for providing water to vegetation.

island   Discrete parcel of land completely surrounded by water. See *islands* under *bar*.

islet   A small island.

isobath   Imaginary line or area on map connecting all points of equal depth.

isocline   Fold or strata that is so tightly compressed that parts of each side dip in the same direction.

isopleth   Line on a graph or chart depicting a measurement of equal value. See *isobath, isotherm*.

isotherm   Line on a chart or graph linking all points having identical water temperatures.

isthmus   Narrow strip of land connecting two larger land areas.

▶ j

jack   See *a-jack* under *habitat enhancements*.

jam   Congested area caused by the accumulation of debris in flowing waters. See also *jam* under *large organic debris*.

jetsam   Materials of human origin that accumulates along coastal areas.

jetty   See *jetty* under *habitat enhancements*.

junction bar   See *junction bar* under *bar*.

▶ k

kame   A ridge-like or hilly local glacial deposit of coarse alluvium formed as a delta at the front of glaciers by meltwater streams.

karst   Geologic features composed of limestone that are characterized by sinks, ravines, and underground streams. Any limestone or dolomitic region with underground geologic features.

k-dam   See *k-dam* under *habitat enhancements*.

kelvin wave   See *kelvin wave* under *wave*.

kettle lake   See *kettle lake* under *lake*.

key dam   Main or primary dam that is generally constructed across the lower part of a gorge.

knickpoint   See *nick point*.

▶ l

lacustrian   Lake-dwelling; term pertains to standing water bodies.

lacustrine   Pertaining to lakes, reservoirs, wetlands, or any standing water body with a total surface area exceeding 8 ha (20 acres). See also *lacustrine* under *streambank material*.

lag deposits   Coarser materials remaining after finer particle-sized materials are removed.

lagoon   (1) Small, shallow, pond-like body of water that is connected to a larger body. (2) Shallow water separated from the sea by a sandbar or a reef. (3) Settling pond for wastewater treatment.

lag time   An interval or lapse in time between events such as precipitation and peak runoff.

lahar   See *lahar* under *landslide*.

lake   (1) Natural or artificial body of fresh or saline water (usually at least 8 ha [20 acres] in surface area) that is completely surrounded by land. (2) Large floodplain feature, persistent, relatively unchanged over a period of years. See *lacustrine*.

   aeolian lake   Lake that forms in a basin scoured out by wind.

   aestival lake   Shallow lake, permanently flooded, that freezes.

alkali lake   Lake where alkaline salts have accumulated from evaporation (i.e., closed basin) or in situations where freshwater inflow is insufficient to dilute the alkaline salts.

bitter lake   Water body formed for salt extraction from seawater or other water body that contains a high concentration of salts in solution.

blind lake   Lake that has neither an inflowing stream nor an outlet. Compare with *seepage lake* under *lake*.

brine lake   Lake with water that is saturated or with high salt concentrations.

caldera lake   Lake formed by subsidence in the roof of a partially emptied magmatic chamber. See *crater lake* and *volcanic lake* under *lake*.

cave-in lake   See *kettle lake, sink lake, thaw lake, themokarst lake* under *lake*.

cirque lake   Lake that occurs where valleys are shaped into structures resembling amphitheaters by the action of freezing and thawing ice. Usually found in the upper portion of a glaciated area or in mountains.

closed lake   Lake that loses water only through evaporation.

coastal lake   Lake formed by the enclosure of lagoons by wave action or an embayment formed by bars deposited by currents along a shoreline.

coldwater lake   Lake with a fish community composed primarily of coldwater fishes.

crater lake   See *caldera lake* and *volcanic lake* under *lake*.

cryogenic lake   Lake formed inside an ice-wedge polygon that develops in permafrost from water seepage through cracks in the surface of the ground.

cutoff lake   Artificial oxbow lake created by river channelization or levee construction that isolates the lake from the river. A cutoff lake may contain a water control structure to facilitate drainage into the river.

dead lake   Body of standing water that cannot support plants or animals because of adverse chemical or physical features.

delta lake   Lake formed upstream of an alluvial deposition.

doline lake   Lake, usually with circular and conically shaped sinks, that is developed from the solution of soluble rock that gradually erodes the rock stratum.

drainage lake   Lake that loses water through flows from an outlet. Sometimes separated into further categories by whether the lake has an inlet and outlet or is spring-fed with no inlet but an outlet.

dune lake   Lake formed by deposition of sand or other material that is moved, primarily or partially, by wind.

dystrophic lake   Lake with a high concentration of humic acid in the water that colors it brown, low pH, and peat-filled margins that develop into peat bogs. See *bog* under *wetlands*.

earthquake lake   Lake formed in a depression created by an earthquake.

ephemeral lake   Lake that contains water for short and irregular periods of time and usually contains water after a period of heavy precipitation.

erosion lake   Lake formed in a depression by the removal or deposition of rock fragments or other organic material.

evorsion lake   Type of pothole formed by streambed abrasion. A category of fluvial lakes associated with torrential flow and cataracts where rocks cut a pothole into the stream bedrock through erosion by a swirling action generated through hydraulic pressure.

extinct lake   A former lake as evidenced by fossil remains or stratigraphy, that no longer holds water due to climatic changes or changes in hydraulic processes.

false karst lake   Lake in a depression caused by geologic piping that results in turbulent subsurface drainage channels in insoluble clastic rock.

finger lake   Lake that is narrow and elongate in shape.

glacial lake   Lake formed by glacial action such as scouring or damming.

graben lake   Lake formed by geologic faulting processes where depressions occur between masses of either a single fault displacement or in multiple downward displacements resulting in troughs. See also *tectonic lake*.

gravel pit lake   Lake formed by water accumulated in a pit from which gravel was excavated below the groundwater level.

intermittent lake   Lake that has surface water only part of the time at different recurring intervals and durations.

kettle lake   Lake formed in a depression by melting ice blocks deposited in glacial drift or in the outwash plain.

landslide lake   Lake formed by dams or depressions created by the mass movement of soil, rock, and debris.

levee lake   Shallow, often elongated body of water that lies parallel to a stream and is separated from it by a strip of higher land.

maar lake   Lake occupying a small depression with a diameter less than 2 km, resulting from lava in contact with groundwater or from degassing of magma.

marl lake   Lake with a high deposition or content of calcareous materials in the littoral zone.

montane freshwater lake   Circumneutral lake found in mountains below timberline.

moraine lake   Lake formed by the damming action of rock debris deposited by retreating glaciers.

open lake   Lake with outflow from an outlet or from seepage.

oxbow lake   Lake occupying a former meander of a river isolated by a shift in the stream channel.

paternoster lake   A chain of small lakes in a glaciated valley formed by ice erosion.

pay lake   Lake where the public pays a fee for swimming, fishing, and other recreational uses.

perched lake   Lake formed by surface water from a perched groundwater table.

perennial lake   Lake that always contains water.

playa lake   An internally drained lake found in a sandy, salty, or muddy flat floor of an arid basin, usually occupied by shallow water only after periods of prolonged heavy precipitation. Sometimes refers to an ephemeral lake that forms a "playa" upon drying up.

plunge lake   An excavated lake at the foot of a waterfall.

pluvial lake   A former lake that existed under a different climate in areas that are now dry.

pothole lake   Ponds, pools, lakes, and wetlands found in depressions (potholes) that were formed by glacial activity.

public lake   Lake legally open to the public for recreational activities that may include boating, fishing, and swimming.

relict lake   A remnant lake or a larger pluvial lake occupying a closed basin in arid regions.

reservoir   A lake formed by impounding water behind a natural or constructed dam.

rift lake   Lake formed along a geologic fault or fault zone.

saline lake   Lake with a high concentration of salts such as carbonates, chlorides, or sulfates.

saltern lake   Lake with a high concentration of sodium chloride.

salt karst lake   Lake formed by the dissolution of evaporites.

seepage lake   Lake without an outlet drained by seepage into groundwater. Compare with *blind lake* under *lake*.

sink lake   Lake that forms in a depression created by underground solution of limestone.

small mountain lake   A small lake (8 ha [20 acres]) formed in a natural topographic depression on a mountain.

soda lake   Lake with a high concentration of $NaHCO_3$ and Na_2CO_3.

solution lake   Lake formed by soluble rock slowly dissolved by percolating groundwater.

stratified lake   Lake with distinctive strata caused by density differences (e.g., temperature, salinity).

tectonic lake   Lake formed in depressions created by movement or damming action of surface plates.

thaw lake   Lake formed in shallow depressions from unequal thawing of permafrost in arctic regions. Also termed *thermokarst*.

thermokarst lake   Lake formed by coalescing of many cryogenic lakes, or by melting of large amounts of ice deep in permafrost.

two-story lake   Lake with an upper layer of warm water supporting warmwater fishes and a lower (i.e., deeper) layer of cold water supporting coldwater fishes.

vernal lake   A shallow lake resulting from seasonal precipitation and runoff that is dry for part of the year.

volcanic lake   Lake formed by depressions in lava or volcanic cones or dams formed by lava that impounds water. See *caldera lake* and *crater lake* under **lake**.

warmwater lake   Lake with a fish community composed primarily of warmwater fishes.

lake volume   See *lake volume* under **dimensions**.

lake width   See *lake width* under **dimensions**.

laminar flow   See *laminar flow* under **flow**.

land   Any part of the earth not covered by water.

landform   Any physical, recognizable form or feature of the earth's surface that is produced by natural processes and has a characteristic or unique shape.

landsat   See *landsat* under **remote sensing**.

landsat multispectral scanner (MSS)   See *landsat multispectral scanner (MSS)* under **remote sensing**.

landsat thematic mapper (TM)   See *landsat thematic mapper (TM)* under **remote sensing**.

landscape   Contiguous heterogeneous natural or constructed land area with similar conditions over a larger geographic area.

landslide   Fall or slide of soil, debris, or rock on or from a slope. Any downslope mass movement of soil, rock, or debris on a landform. See also *landslide pool* under *slow water, pool, dammed pool* under the main heading **channel unit**.

avalanche   Large mass of snow or ice, sometimes with rocks and vegetative debris that moves rapidly downslope.

avalanche cone   Accumulated material, similar to a talus cone, deposited by a snow avalanche.

avalanche track   Channel-like pathway of an avalanche, marked by eroded surfaces and bent or broken trees.

bedrock landslide   Infrequent landslides of rock whose strength is much higher than the strength of their boundaries. These landslides usually occur as frequent small slides that produce irregular hillslopes with steep toes and head scarps.

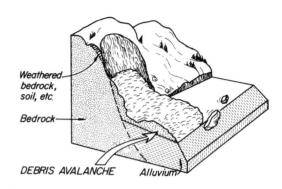

debris avalanche (from Meehan 1991)

debris avalanche   Rapid, shallow, downslope mass movement of saturated soil or surficial material (commonly including vegetation debris) that occurs on steep hillsides.

debris fall   Fall or rolling soil or surficial material mass.

debris flow   Rapid mass movement of material including vegetation, soil, mud, boulders, and weathered rock (i.e., a debris torrent) down a steep (greater than 5°) mountain slope or stream channel. See **debris torrent**.

debris flow tracks   Erosion marks made by debris flows characterized by the lack of vegetation or young plants, deposits of soil, boulders, and rock, levees, or gullies.

debris jam or dam   Accumulation of debris in a channel that may cause ponding of water or alluvial deposition upstream from the accumulation, and block fish movements.

debris slide   Downslope sliding of a mass of soil or surface material that may be transformed into a debris flow.

debris torrent   Rapid, turbulent movement of water, soil, alluvium, and organic matter down a stream channel during storms or floods. Torrents generally occur in small streams with scouring of the streambed and deposition of a large quantity of material at the lower end of the torrent. See **debris flow**.

deep-seated creep Slow, gradual, more or less continuous and permanent deformation of soil caused by gravity.

deep-seated failures Landslides involving deep regolith, weathered rock, or bedrock, as well as surface soil that may form large geologic features.

dry ravel Downslope movement of dry, noncohesive soil or rock particles by gravity in a form of soil creep.

earthflow Mass downslope movement of soil and particles of weathered rock smaller than sand that results when soils become saturated and unstable.

earth slump Downslope movement of a relatively intact mass of soil, colluvium, and vegetation along a clearly defined, concave surface that is parallel to the slope.

flow slides Rapid downslope movement of a cohesive soil mass that results from liquefaction of a bank in saturated silty and sandy soils.

flow till Debris flow resulting from melting ice that causes saturation of glacial debris and accumulates on a lower gradient, more stable surface.

gravity flow Downslope flow of a mixture of water and sediment caused by slumping.

headscarp Steep, upper portion of a landslide scar.

lahar Flash floods or avalanche consisting of semiliquid mud, rock, and ice that surge rapidly downslope from higher altitudes, and usually occur on the slopes of volcanoes.

landslide scar Part of a slope (commonly steep) that is exposed or visibly modified from downslope movement of soils and rock accompanying a landslide.

large persistent deep-seated failure Slump-earthflows involving large areas of hillside on landscapes that remain recognizable for a long period of time.

mass movement General term for a large downslope movement of soil and rock or for avalanches that result from natural or anthropogenic causes.

mass wasting Downslope movement, flow, or slumping of a large amount of coherent or aggregate material from weathering and erosion that reduces the steepness of slopes and deposits the material in lower areas.

mud slide Horizontal and vertical movement of mud caused by natural water saturation of the soil or from anthropogenic disturbance.

periglacial landslide Active valley wall processes characteristic of very cold regions such as next to active glaciers or alpine terrain.

rock avalanche Rapid, downslope flow of large masses of rock by rock falls and slides from bedrock areas.

rock creep Slow, gradual downslope progression of rock fragments.

rock fall Relatively free-falling and precipitous movement of newly detached fragments of bedrock (any size) from a cliff or other very steep slope.

rock slide Downslope movement of rock on or from a steep slope.

rotational failure Horizontal downslope movement of hillside material that results in an upward turn on the toe of the slide. See *slump*.

shallow-rapid landslide Shallow (1–2 m) landslide on a steep slope produced by failure (slipping, sliding) of the soil mantle including glacial till and some weathered bedrock that leaves an elongate, spoon-shaped scar.

slab failure Broad, flat, somewhat thick rock that breaks off or falls as a single unit.

slide Mass movement of soil and rock where slope failure occurs along one or more slip surfaces. May become a debris flow or torrent flow if enough water is present in the mass.

slip erosion Downslope slide of a whole mantle rock. The large, spectacular forms are termed landslips, landslides, or debris avalanches.

slope failure Rupture and collapse or flow of surface materials, soil, or bedrock due to shear stress exceeding the strength of the material.

slump Deep, rotational landslide, generally producing coherent lateral and downward movement (back-rotation) of a mass of soil, rock, or other material that slips as a coherent

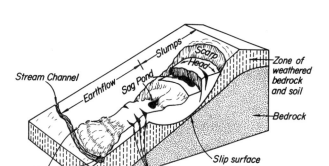

slump (from Meehan 1991)

mass. Also, a land area that sunk to create a boggy, marshy place.

slump–earthflow   Landslide exhibiting characteristics of slumps and earthflows in which the upper part moves by slumping while the lower portion moves by flow.

small sporadic deep-seated failures   Commonly smaller slumps that can result from storms or earth movement at irregular time intervals and can crumble or decay so that they are indiscernible.

snow avalanche   See *avalanche* under *landslide*.

landslide lake   See *landslide lake* under *lake*.

landslide pool   See *landslide pool* under *slow water*, *pool*, under the main heading **channel unit**.

landslide scar   See *landslide scar* under *landslide*.

landtype   Unit on the earth's surface of a characteritistic geomorphic surface type and a particular lithological composition, indentifiable on a spatial scale of hectares or acres.

langmuir circulation   See *langmuir circulation* under *wave*.

large benthic organic matter   See *large benthic organic matter* under **benthic organic matter**.

large bole   See *large bole* under *large organic debris*.

large impoundment   An impoundment larger than 227 hectares (500 acres) in surface area.

large organic debris   Large woody material (e.g., log or tree) with a diameter greater than 10 cm (4 in) and a length greater than 1 m (39 in).

Synonymous with large woody debris (LWD), a term that is commonly used in the Pacific Northwest where organic debris generally is a log or tree.

aggregate   Two or more large woody pieces at one location.

biological legacies   Large trees, logs, snags, and other components of a forest that remain after logging for reseeding and that provide terrestrial or aquatic ecological structure.

blowdown   Trees that have fallen from high wind.

bole   Term referring to the trunk of a tree.

clump   Irregular accumulation of debris along a stream channel that does not form major impediments to streamflow.

deadhead   Log submerged and close to the water surface that is not embedded, lodged, or rooted in the stream channel.

digger log   Log anchored to a streambank or channel bottom to form a scour pool.

down log   Portion of a tree that has fallen or been cut and left on the forest floor.

fixed logs   Log (or group of logs) that is firmly embedded, lodged, or rooted in a stream channel.

floats   Accumulations of logs floating on the surface but prevented from moving downstream by an obstruction.

free log   Log (or group of logs) that is not embedded, lodged, or rooted in the stream channel.

jam   Wholly or partially submerged accumulation of woody debris from winds, water currents, or logging activities that partially or completely blocks a stream channel and obstructs streamflow.

large bole   Bole 60 cm (24 in) or more in diameter.

log   In general, a tree trunk, bole, or large limb, with or without the roots attached.

log dam   Dam formed by the deposition of woody debris and stream sediments at an obstruction caused by a log or several logs in the stream channel.

root wad   Root mass from a tree. Synonymous with butt ends.

scattered   Single pieces of debris distributed at irregular intervals along a stream channel.

small bole   Bole less than 60 cm (24 in) in diameter.

snag   Generally applies to a standing dead tree. Sometimes applied to a submerged fallen tree in a large stream with the top exposed or only slightly submerged.

stable debris   Large woody debris, usually anchored or embedded in the stream bottom or bank that is not dislodged during periods of high flows.

sweeper log   Fallen tree with branches that accumulates floating logs and other debris and projects into the channel, creating a hazard for navigation.

volume   Volume (V) of a piece of large woody debris can be calculated as:

$$V = \frac{\pi(D_1^2 + D_2^2)L}{8} \; ;$$

D_1 = diameter of debris at one end;
D_2 = diameter of debris at other end;
L = length.

woody debris   Collection of materials in the water or substrate on the bank or shoreline that is primarily composed of wood.

large persistent deep-seated failure   See *large persistent deep-seated failure* under **landslide**.

lateral channel movement   Movement of a stream channel laterally across a valley floor either gradually through meandering or suddenly through avulsions.

lateral moraine   See *lateral moraine* under **moraine**.

lateral scour pool   See *lateral scour pool* under *slow water, pool, scour pool* under the main heading **channel unit**.

late succession stage   See *late succession stage* under **succession**.

leachate   Soluble substance that has been removed from other material by water percolation.

leaching   Removal of soluble material in the ground by percolating water.

leak   Escape or entry of water.

leakage   The amount of water that leaks into or out of a storage area.

leaking   Movement of water into or out of a storage area or stream.

least flow   See *least flow* under **flow**.

leave strips   Bands of trees or other vegetation left along stream, rivers, and roads as buffers from adjacent forest management and other human activities. See also **buffer**.

ledge   Reef, ridge, shelf, or line of rock longer than wide found in the sea, lakes, or other bodies of water.

lee   The downwind side of an object being affected by some physical process.

lee bar   See *lee bar* under **bar**.

leeshore   The shore protected from direct wind or a shore receiving wind only from the land.

leg-type structure   See *leg-type structure* under **habitat enhancements**.

length   See *length* under **dimensions** and **wave**.

lentic   An aquatic system with standing or slow flowing water (e.g., lake, pond, reservoir, swamp, marsh, and wetland). Such systems have a nondirectional net flow of water. See **standing water**.

levee   (1) Naturally formed elongate ridge or embankment of fluvial sediments deposited along a stream channel. (2) Artificial embankment along a water course or an arm of the sea to prevent flooding or restrict movement of water into or through an area. Compare with *berm* under **habitat enhancements, dike**.

levee lake   See *levee lake* under **lake**.

level   Surface elevation of a body of water relative to mean sea level.

Liebig's law of the minimum   Growth and reproduction of an organism is hindered when a minimum quantity of an element or substance such as oxygen, carbon dioxide, or calcium is available.

limiting factor (1) Any condition that approaches or exceeds the limits of tolerance by an organism. (2) A habitat component (biological, chemical, or physical) whose quantity constraints or limits the size of a population.

limnetic (1) Open-water zone of a water body too deep to support rooted aquatic vegetation. (2) Zone of deep water between the surface and compensation depth.

limnic material Soils made up of inorganic or organic compounds that are deposited in water by precipitation, aquatic organisms, underwater or floating aquatic plants, or aquatic animals.

limnocrene A spring pool (generally large) with or without outlet.

linear extent of structure Percentage of sides in a water body that are cliffs or shoals.

link In drainage networks, stream reach of a particular order, or a stream at its origin.

  downstream link Magnitude of a link at the next downstream confluence.

  link magnitude In drainage networks, the number of links upstream from a given point in the network.

lithology Physical character of a rock or deposit that is defined by rock type or distribution of particle sizes.

lithorheophilic See *lithorheophilic* under **benthos**.

litter Leaves, needles, twigs and branches, and other small organic matter entering a water body. Also referred to as litter fall.

littoral Shallow shore area (less than 6 m [~20 ft] deep) of a water body where light can usually penetrate to the bottom and that is often occupied by rooted macrophytes.

  epilittoral Shore area entirely above the water line that is unaffected by spray.

  eulittoral Shoreline between the highest and lowest seasonal water levels.

  littoroprofundal Zone below the lower edge of the macrophyte boundary.

  lower infralittoral Zone of submerged vegetation.

  middle infralittoral Zone of floating leaf vegetation.

  sublittoral Portion of a shore from the lowest water level to the lower boundary of plant growth.

  supralittoral Shore area above the water line that is subjected to spraying by waves.

  upper infralittoral Zone of rooted vegetation.

littoral drift Material floating in the littoral zone of a lake.

littoral sediment Sediment deposited along the shore of a lake.

littoroprofundal See *littoroprofundal* under **littoral**.

loading Addition of specific quantities of substances or heat to a water body.

loess Fine, loamy soil or sand deposited by the wind.

log See *log* under **large organic debris**.

log dam See *log dam* under **large organic debris**.

log pond See *log pond* under **pond**.

log sill See *check dam* under **habitat enhancements**.

long-duration flooding See *long-duration flooding* under **flooding**.

longitudinal profile A plot of elevations with distances to depict stream channel characteristics.

longshore drift Movement of sediment along a beach from wash and backwash of waves that make oblique contact with the shore.

long wave See *long wave* under **wave**.

losing stream See *losing stream* under **stream**.

lotic Aquatic system with rapidly flowing water such as a brook, stream, or river where the net flow of water is unidirectional from the headwaters to the mouth.

lotic (riparian) wetland See *lotic (riparian) wetland* under **wetlands**.

lower bank See *lower bank* under **streambank**.

lower infralittoral   See *lower infralittoral* under *littoral*.

lower valley wall tributary   See *lower valley wall tributary* under **tributary**.

low flow   See *low flow* under **flow**.

low gradient riffle   See *low gradient riffle* under *fast water—turbulent, riffles* under the main heading **channel unit**.

low-head dam   A low barrier that is placed in a waterway to retain or redirect flows. See also *check dam* under **habitat enhancements**.

lowland   Land that is level and only slightly above the elevation of the water surface and that is periodically subject to flooding.

lowland reservoir   Impoundment constructed in rolling hill country or on a stream with a moderate floodplain.

lowland stream   See *lowland stream* under **stream**.

low moor   See *low moor* under **wetlands**.

lunker structure   See *lunker structure* under **habitat enhancements**.

▶ m

maar lake   See *maar lake* under **lake**.

macroinvertebrate   An invertebrate animal (without backbone) large enough to be seen without magnification and retained by a 0.595 mm (U.S. #30) screen. See *benthos*.

  collectors   Macroinvertebrates that feed on living algae, fine particulate matter, or decomposing particulate organic matter collected by filtering water or the surface of sediments.

  predators   Macroinvertebrates that feed on flesh or fluids of other animals.

  scrapers   Macroinvertebrates that feed on algae or microflora attached to the substrate or to plants.

  shredders   Macroinvertebrate herbivores and detritivores that feed on living and decomposing vascular plant tissue.

macroinvertebrate community indicators   See *water quality indicators*.

macrophyte   A plant that can be seen without the aid of optics.

  emergent macrophyte   Aquatic plants growing from submerged soils in water between 0.5 and 1.5 m (1.5–5 ft) deep.

  floating macrophyte   Floating aquatic vegetation growing from submerged substrates in water between 0.5 and 3.0 m (1.5–10 ft) deep.

  free-floating macrophyte   Aquatic vegetation at or near the surface not rooted in the substrate.

  nonpersistent emergent   Emergent hydrophytes whose leaves and stems die and decompose annually so that most portions of the plant above ground are easily transported by currents, waves, or ice.

  rheophyte   A plant associated with fast-flowing water.

  submerged macrophyte   General term for aquatic vegetation in the photic zone.

macroplankton   See *macroplankton* under **plankton**.

madicolous habitat   Thin sheets of water flowing over rock faces, found at the edge of rocky streams, at the sides of waterfalls, and on rocky chutes.

main channel   See *main channel* under **channel pattern**. Also referred to as *main stem channel* under **channel pattern**.

main channel pool   See *main channel pool* under *pool, scour pool* under the main heading **channel unit**.

main-lake point   A peninsula that juts into the main body of a lake or reservoir and continues underwater.

main stem   Principal, largest, or dominating stream or channel in any given area or drainage system.

major waterway   Any river, stream, or lake that is extensively used by commercial or private watercraft.

mangrove swamp   See *mangrove swamp* under **wetlands**.

Manning's *n* An empirical coefficient for computing stream bottom roughness. Used in determining water velocity in stream discharge calculations. In English units,

$$Q = \frac{1.486 R^{2/3} S^{1/2} A}{n} \ ;$$

Q = discharge;
R = hydraulic radius;
S = energy gradient (parallel to water slope);
A = cross-sectional area.

Because *Q* = water velocity, (*V*) times area (*A*),

$$V = \frac{1.486 R^{2/3} S^{1/2}}{n} \ .$$

In metric units, the coefficient 1.486 becomes 1.0.

marginal habitat Habitats that approach the limits to which a species is adapted. More generally, habitats with physical or biological factors that support only limited species or populations.

marina Area to dock boats and provide other services for small watercraft.

marine Of, or pertaining to, the ocean and associated seas. See also *marine* under **streambank material**.

marl Light gray, calcareous, generally friable, clay or clay-loam composed principally of carbonate derived from photosynthetic activity of algae and mollusk shells.

marl lake See *marl lake* under **lake**.

marsh See *marsh* under **wetlands**.

mass movement See *mass movement* under **landslide**.

mass wasting See *mass wasting* under **landslide**.

mat (1) Collection of floating debris, macrophytes, or algae. (2) Submerged artificial fish structure, particularly one made of brush and tree branches.

mature river See *mature river* under **river**.

mature stream See *graded stream* under **stream**.

maximum bank height See *maximum bank height* under **dimensions**.

maximum depth See *maximum depth* under **dimensions**.

maximum elevation See *maximum elevation* under **dimensions**.

maximum flow See *maximum flow* under **flow**.

maximum lake length See *maximum lake length* under **dimensions**.

maximum likelihood See *maximum likelihood* under **remote sensing**.

maximum wavelength See *maximum wavelength* under **wave**.

maximum width See *maximum width* under **dimensions**.

mean annual flow See *mean annual flow* under **flow**.

mean annual runoff See *mean annual runoff* under **flow**.

mean catchment slope Can be calculated by:

$$S_b = \frac{(\text{elevation at } 0.85L) - (\text{elevation at } 0.10L)}{0.75L} \ ;$$

L = maximum length of the catchment basin.

mean column velocity See *mean column velocity* under **velocity**.

mean cross section velocity See *mean cross section velocity* under **velocity**.

mean density of structural units Density of structures per unit of shoreline or bottom.

meander Sinuous course of a river having specific geometric dimensions that describe the degree of curvature. More particularly, one curved portion of a sinuous or winding stream channel, consisting of two consecutive loops, one turning clockwise and the other counterclockwise. See **sinuosity**.

acute meander Meander with sharp, hairpin turns.

amplitude Breadth or width of the meander.

belt of meander The approximate width of a stream valley.

compound meander Irregular meander pattern developed on streams with more than one dominant discharge.

confined meander Meander scrolls within a confined floodplain.

distorted meander Meanders where obstructions limit lateral movement and development of sinuosity.

flat meander An irregular meander in a stream with a flat streambed.

free meander Meander in unconsolidated alluvium that migrates freely to develop waveforms without constraint from valley walls, adjacent terrain, or distortion from heterogeneous alluvium.

irregular acute meander Meander with an irregular pattern of sharp hairpin turns.

irregular meander Deformed or irregular meanders of variable size within a meander belt.

irregular meander with oxbow Irregular meander that retained a remnant loop.

meander belt-width Normal width or distance between tangents drawn on the convex sides of successive belts.

meander length Wave length of the meander.

meander scar Evidence of old channel locations from the lateral migration of a meander.

paleopotamon A large, deep former river meander that is disconnected from the current river channel except at the highest flood flow.

parapotamon Dead arms that are permanently connected to a river channel at the downstream end.

plesiopotamon A former braided channel connected only during high flows.

point of inflection Location where the thalweg crosses from one bank to the other.

radius of curvature Degree of curvature of the meander loop.

regular meander Meander with regular intervals and a consistent pattern of sinuosity and loops.

simple meander Meander with one dominant meander belt-width and wavelength.

sinuous meander Meander with slight curvature within a stream reach of less than approximately two channel widths.

straight meander Meander with very little curvature.

tortuous meander Meander that has a more or less repeated pattern characterized by angles greater than 90°.

truncated meander Meander where confinement limits lateral movement and sinuosity with undeveloped loops.

unconfined meander Meander that migrates unrestricted across a floodplain.

meander belt-width See *meander belt-width* under *meander*.

meander length See *meander length* under *meander*.

meander line A survey line that represents the location of the actual shoreline of a permanent natural body of water, without showing all the details of its windings and irregularities.

meander scar See *meander scar* under *meander*.

meander scrolls channel See *meander scrolls channel* under *channel pattern*.

mean flow See *mean flow* under *flow*.

mean high water Average height of the high water over 19 years.

mean lake depth See *mean lake depth* under *dimensions*.

mean lake width See *mean lake width* under *dimensions*.

mean low water Average height of the low water over 19 years.

mean monthly discharge See *mean monthly discharge* under *discharge*.

mean sea level Average elevation of the surface of the sea for all tidal levels over 19 years. Altitudes or elevations are expressed as elevations above mean sea level.

mean stream length See *mean stream length* under *dimensions*.

mean stream slope The mean gradient or drop in stream surface elevation per unit length of stream.

$$S_c = \frac{\text{elevation at source} - \text{elevation at mouth}}{\text{length of stream}} .$$

medial moraine See *medial moraine* under **moraine**.

median depth See *median depth* under **dimensions**.

median lethal dose (LD50 or LD$_{50}$) Dose (internal amount) of a substance that is lethal to 50% of a group of organisms within a specified time period.

median lethal concentration (LC50 or LC50) The concentration of a substance that kills half of the test organisms in a specified period of exposure. Called the *median tolerance limit* in older literature.

median line Line that is equidistant from the nearest points on opposite shores that forms the center line of the channel.

median moraine See *median moraine* under **moraine**.

median tolerance limit (TLm or TL$_m$) See *median lethal concentration*.

median width See *median width* under **dimensions**.

melt Action by which a solid becomes a liquid.

meltout till See *meltout till* under **moraine**.

meltwater Water that originates from the melting of snow or ice.

meltwater channel See *meltwater channel* under **channel pattern**.

meltwater stream See *meltwater stream* under **stream**.

merge See *merge* under **remote sensing**.

meromictic See *meromictic* under **mixing**.

mesa High, nearly flat-topped and usually isolated area bounded by steep slopes.

mesic Pertaining to or adapted to an area that has a balanced supply of water (i.e., moderately wet).

mesocline See *mesocline* under **stratification**.

mesohaline See *mesohaline* under **salinity**.

mesolimnion See *mesolimnion* under **stratification**.

mesophytic Plant species that grow in locations where soil moisture and aeration conditions lie between extremes (i.e., occurring in habitats with average moisture conditions).

mesosaline See *mesosaline* under **salinity**.

mesotrophic See *mesotrophic* under **trophic**.

metalimnion See *metalimnion* under **stratification**.

metaphyton Algae aggregated in the littoral zone, neither attached to substrate nor planktonic.

metapotamon See *metapotamon* under **potamon**.

metarhithron See *metarhithron* under **rhithron**.

meteoric water Term applied to rainwater, snow, hail, and sleet.

microclimate The climatic factors operating in a terrestrial microenvironment (i.e., small or restricted area).

microenvironment All the external environmental conditions that may influence an organism's physiology or energetics in a small or restricted area.

microhabitat Specific locations where organisms live that contain combinations of habitat characteristics that directly influence the organisms at any life stage. See **niche**.

microplankton See *microplankton* under **plankton**.

mid-channel bar See *mid-channel bar* under **bar**.

mid-channel pool See *mid-channel pool* under *pool, scour pool* with the main heading of **channel unit**.

middle infralittoral See *middle infralittoral* under **littoral**.

middle stream See *middle stream* under **stream**.

mid-successional stage See *mid-successional stage* under *succession*.

milk White or chalky-colored appearance of suspended mineral particles in water.

mill pond See *mill pond* under **pond**.

mine discharge   Water that is drained, pumped, or siphoned from a mine.

mineral spring   Spring with water containing a significant amount of dissolved minerals.

mineral turbidity   Turbidity resulting from the presence of fine mineral particles such as clay.

mineral water   Water containing a significant amount of dissolved minerals.

minimum bank height   See *minimum bank height* under *dimensions*.

minimum flow   See *minimum flow* under *flow*.

minor discharge   Any discharge of less than 50,000 gallons per day or a discharge that does not adversely affect the receiving waters.

mire   Refers to slimy soil or deep mud but also may be applied to swampy ground, bogs, or marshes.

mitigation   (1) Action taken to alleviate or compensate for potentially adverse effects on aquatic habitat that have been modified through anthropogenic actions. (2) In-kind mitigation may be substituted for compensation to replace a resource that has been negatively impacted with a similar resource (e.g., a stream for a stream). (3) Out-of-kind mitigation refers to replacement of one resource with another (e.g., a lake for a stream).

mixing   Internal circulation in a water body.

  amictic   Lakes that do not mix because they are perennially sealed off by ice from most of the annual variation in temperature.

  biogenic meromixis   Lakes that do not mix because of salt accumulation in the hypolimnion from decomposition of organic matter in the sediment.

  cold monomictic   Lakes with water temperature never greater than 4°C that circulate only one time in summer at or below 4°C.

  crenogenic meromixis   Lakes that do not mix because of high salinity from saline springs at deep pockets in the basin.

  dimictic   Lakes with two periods of mixing—one in the spring and one in the fall.

  ectogenic meromixis   Lakes that do not mix because of density differences. See *meromictic* under *mixing*.

  holomictic   Lakes that are mixed throughout the water column by wind.

  interflow   Flow of water into a lake that is greater in density than the epilimnion or hypolimnion and remains as a plume at intermediate depths.

  meromictic   Lake in which dissolved substances create a density gradient with depth that prevents complete (top to bottom) mixing or circulation of water. Periodic circulation occurs only in waters above the chemocline.

  mixing characteristics   Frequency with which all or some part of water is mixed.

  mixing depth   Depth of body of water where mixing occurs.

  mixing zone   Area or location of a water body where individual masses of water are mixed.

  monomictic   Lake with one regular period of circulation per year.

  oligomictic   Lake with temperatures always greater than 4°C that exhibit only irregular circulation.

  overflow   The inflow of water at the surface of a lake that is less dense than the lake water.

  partial meromixis   A normally dimictic lake that skips a turnover, usually in spring, from dynamic processes such as decomposition.

  polymictic   Lake with frequent or continuous circulation.

  turnover ratio   Volume of water (as a percentage) that is affected by turnover in meromictic lakes.

  underflow   Flow of water into a lake that has a greater density than the water in the lake.

  warm monomictic   Waters that do not drop below 4°C and circulate freely in winter but stratify in summer.

  warm polymictic   Waters that circulate freely at temperatures well above 4°C.

mixing characteristics   See *mixing characteristics* under *mixing*.

mixing depth   See *mixing depth* under *mixing*.

mixing zone   See *mixing zone* under *mixing*.

mixohaline   See *mixohaline* under *salinity*.

mixolimnion   See *mixolimnion* under *stratification*.

mixosaline   See *mixosaline* under *salinity*.

modified flow   See *modified flow* under *flow*.

modified universal soil loss equation   An estimate of sediment yield for an individual storm event can be calculated from:

$$Y = 95(Qq)^{0.56}KLSCP \; ;$$

Y = single storm sediment yield, in tons;
Q = storm runoff in acre-feet;
q = peak discharge in cubic feet per second;
K = soil erodibility factor;
L = slope length factor;
S = slope steepness factor;
C = cover management factor;
P = supportive practices factor.

monimolimnion   See *monimolimnion* under *stratification*.

monomictic   See *monomictic* under *mixing*.

monsoon   (1) Wind that affects the climate of a large area when it changes direction with the seasons. (2) Specifically applies to the seasonal wind in southern Asia that blows from southwest in summer and northeast in winter and is associated with episodes of heavy rainfall separated by periods of little or no rain.

montane   Pertaining to mountains or mountainous areas.

montane freshwater lake   See *montane freshwater lake* under *lake*.

monthly mean discharge   See *monthly mean discharge* under *discharge*.

moor   See *moor* under *wetlands*.

morainal   See *morainal* under *streambank material*.

moraine   Irregular, surficial geologic deposit of sand, rock, and debris left by a retreating glacier.

ablation moraine   Moraine formed by melting ice that has a typically hummock form.

end moraine   A ridge of glacial till that remains in equilibrium at the terminus of a valley glacier or at the margin of an ice sheet.

ground moraine   Thin deposits left underneath a retreating glacier that may have a gently rolling or hummock-like appearance.

hummocky moraine   Moraine composed of generally random knobs, ridges, hummocks, and depressions.

ice-disintegration moraine   Moraine formed by accumulation of till and other materials on toe of stagnant ice.

lateral moraine   Geologic deposits formed along the sides of a retreating glacier.

medial moraine   A long strip of rock debris carried on or within a glacier from the convergence of lateral moraines where two glaciers join.

median moraine   Geologic deposits along the central path of a glacier.

meltout till   Material that collects on or under glacier ice of a stationary or stagnant glacier.

neoglacial moraine   Term applied to a moraine that formed during the Neoglacial period and the more recent Little Ice Age.

recessional moraine   Moraine created by accumulation of materials deposited by a melting glacier that marks a temporary halt in its general retreat.

terminal moraine   Geologic deposits at the front lobe or foot of a glacier that marks the furthest point reached by a glacier.

moraine lake   See *moraine lake* under *lake*.

morass   See *morass* under *wetlands*.

morphoedaphic index (MEI)   An index of the trophic state of a water body:

$$MEI = [(TDS)/\bar{d}\,]^{\frac{1}{2}} \; ;$$

TDS = total dissolved solids;
d = mean depth.

morphology   Physical attributes of a water body and the methods for measuring those attributes.

morphometry   The physical shape of a water body, such as a stream, lake, or reservoir.

mosaic See *mosaic* under ***remote sensing***.

moss–lichen wetland See *moss–lichen wetland* under ***wetlands***.

mountain processes See *mountain processes* under ***active valley wall processes***.

mouth Downstream terminus of a stream as it enters another water body.

moveable bed A streambed composed of materials that are readily transported by streamflow.

muck Soft fine-grained soil composed of silt, clay, or organic substrate material, typically dark in color, that consists of 20–50% highly decomposed organic matter with intermingled silt and clay.

mud (1) Wet, sticky earth composed of silt intermingled with clay that may contain organic material. (2) Term that is often applied to firm streambeds composed of soil.

mudbank Lateral sides of a streambank created by mud deposition.

mud cracks Cracks formed in mud as it dries and shrinks.

mud flats Shallow areas of a stream composed of silt and other fine particles that are periodically exposed at relatively even elevations.

mudflow Lateral flow of mud that has been wetted by precipitation.

mudsill See *mudsill* under ***habitat enhancements***.

mud slide See *mud slide* under ***landslide***.

multipurpose reservoir Artificial impoundment used for water storage where water releases are regulated for various uses including domestic water supply, irrigated agriculture, power generation, and navigation.

multispectral classification See *multispectral classification* under ***remote sensing***.

multispectral imagery See *multispectral imagery* under ***remote sensing***.

multispectral scanner (MSS) See *multispectral scanner (MSS)* under ***remote sensing***.

muskeg See *muskeg* under ***wetlands***.

▶ n

nadir See *nadir* under ***remote sensing***.

nannoplankton Planktonic organisms that are small enough to pass through a 0.03 mm mesh net. See also *nannoplankton* under ***plankton***.

ultranannoplankton Plankton that are less than 0.2 μm.

nannoseston See *nannoseston* under ***seston***.

nasmode A spring complex or an area where a number of nearby springs originate from the same groundwater source.

native species Plant and animal species that occur naturally in aquatic and terrestrial habitats.

natural erosion See *natural erosion* under ***erosion***.

natural flow See *natural flow* under ***flow***.

natural levee Sediment deposited on a streambank as floodwaters subside that creates natural banks or ridges slightly above the floodplain.

natural spillway See *natural spillway* under ***control structure***.

navigable Waterways used by humans for transportation or transport of goods. Legally applied to interstate waters: intrastate lakes, rivers, and streams that are used by interstate travelers for recreational or other purposes; intrastate lakes, rivers, and streams from which fish or shellfish are taken and sold in interstate commerce; and intrastate lakes, rivers, and streams that are used for industrial purposes by industries involved in interstate commerce.

nearshore Zone extending from a shore to a distance where the water column is no longer influenced by conditions on or drainage from land.

neck Narrow strip of land, such as an isthmus, cape, or strait.

needle ice Thin ice crystals formed on soil and rocks from frost-heaving.

negative heterograde See *negative heterograde* under ***stratification***.

nekton   Actively swimming organisms able to move independent of water current.

neoglacial moraine   See *neoglacial moraine* under *moraine*.

neritic zone   Relatively shallow water zone in oceans or seas that extends from the high-tide mark to the edge of the continental shelf.

net primary production   See *net primary production* under *production*.

net seston   See *net seston* under *seston*.

neuston   Microscopic organisms associated with habitats at the interface of air and water. See *pleuston*.

   epineuston   Neuston living on the upper surface of water.

   hyponeuston   Neuston living on the underside of the surface water film.

neutral estuary   Type of estuary where neither the freshwater inflow nor evaporation predominates (i.e., total freshwater inflow and precipitation equals evaporation).

niche   (1) Ecological position of an organism within its community or ecosystem that results from the organism's structural adaptations, physiological responses, and specific behavior. An analogy is, the habitat is an organism's "address," while its niche is its "profession." (2) Ecological and functional role of an organism in a community, especially with regard to its food consumption. See *microhabitat*.

nick point (knickpoint)   (1) Narrowing of a channel causing an increase in current velocity that results in an upstream accumulation of water and deposition. (2) Abrupt changes in slope at the confluence of streams or associated with geologic features.

nodal point   See *nodal point* under *wave*.

node   Refers to the ending points of a line that is used in GIS systems as a reference point along a stream.

nonclastic   Crystalline chemical precipitates forming sedimentary deposits.

nonflooding floodplain   See *nonflooding floodplain* under *floodplain*.

nonfoliar shading   See *nonfoliar shading* under *stream surface shading*.

nonhydric   Soils that develop under predominantly aerobic soil conditions.

nonmarine sediment   Sediment that accumulates in rivers or freshwater lakes.

nonpersistent emergent   See *nonpersistent emergent* under *macrophyte*.

nonpersistent wetland   See *nonpersistent wetland* under *wetlands*.

nonpoint source   Usually applied to pollutants entering a water body in a diffuse pattern rather than from a specific, single location that includes land runoff, precipitation, atmospheric deposition, or percolation.

nonsaturated zone   Zone where the actual amount of oxygen or other material dissolved in water is less than saturation.

nonsensitive reservoir   Reservoir where overall productivity is not decreased by low reservoir production rates.

nonwetland   Area that is sufficiently dry so that hydrophytic vegetation, hydric soils, and wetland hydrology do not occur; includes uplands as well as former wetlands that are drained.

nonwoody benthic organic matter   See *nonwoody benthic organic matter* under **benthic organic matter**.

normal erosion   See *normal erosion* under *erosion*.

normal high water   High water mark that occurs annually in a water body. In streams, it occurs at bank-full flows and is also called the peak annual flow (QFP).

nose velocity   Water velocity immediately in front of a fish that is positioned into the current.

nuisance organism   Term applied to an organism that is capable of interfering with the use or treatment of water.

nutrient   Element or compound essential for growth, development, and life for living organisms such as oxygen, nitrogen, phosphorus, and potassium.

nutrient budget   Gain and loss of all nutrients in a specific water body.

nutrient cycling   Circulation of nutrient elements and compounds in and among the atmosphere, soil, parent rock, flora, and fauna in a given area such as a water body.

nutrient depletion   Situation where the export of nutrients is greater than the import and where the reduction in the total amount of nutrients and their rate of uptake, release, movement, transformation, or export negatively affects organisms that inhabit a particular area.

nutrient loading   Addition of nutrients into the water column via runoff, discharge, internal recirculation, groundwater, or atmosphere.

nutrient spiraling   Cycling and downstream transport of nutrients from physical and biological activities in a stream.

▶ **O**

oasis   Isolated, fertile area with vegetation in an arid region that is supplied with water from a well or spring.

obligate wetland species   See *obligate wetland species* under *wetland status*.

obsequent stream   See *obsequent stream* under *stream*.

obstruction   Object or formation that partially or wholly blocks or hinders water flow and movement of organisms or that restricts, endangers, or interferes with navigation. Examples in aquatic habitats include geologic features, falls, cascades, chutes, beaver dams, and dams on impoundments.

occasionally flooded floodplain   See *occasionally flooded floodplain* under *floodplain*.

oceanadromous   Life cycle strategy of fish including migrations, reproduction, and feeding that occur entirely in saltwater. Compare with *anadromous, catadromous, diadromous, and potamodromous*.

off-channel pond   See *off-channel pond* under *pond*.

off-channel pools   See *off-channel pools* under *habitat enhancements*.

offshore   Away from, moving away from, or at a distance from the shore.

offshore wind   Wind blowing away from the shore.

ogee   See *ogee* under *control structure*.

old river   See *old river* under *river*.

oligohaline   See *oligohaline* under *salinity*.

oligomictic   See *oligomictic* under *mixing*.

oligopelic   Bottom deposit containing very little clay.

oligosaline   See *oligosaline* under *salinity*.

oligosaprobic zone   See *oligosaprobic zone* under *saprobein system*.

oligotrophic   See *oligotrophic* under *trophic*.

onshore   In water, refers to a location on, moving onto, close to, or parallel to the shore. On land, refers to a location adjacent to a water body.

onshore wind   Wind blowing toward the shore.

ooze   (1) Soft, fine-textured bottom mud. (2) A condition where a substance flows, or is extruded, very slowly through openings.

opacity   Degree of obstruction to light passage.

open basin   Basin with a surface outlet.

open lake   See *open lake* under *lake*.

open water   (1) Water free of vegetation, stumps, or artificial obstructions that is away from the shoreline. (2) Water in a pond, lake, or reservoir that remains unfrozen or is not covered by ice during winter.

operating pool   Amount of water retained in a reservoir for limited periods and released to operate turbines.

optimum flow   See *optimum flow* under *flow*.

optimum level   The most suitable degree of any environmental factor for the well-being, health, and productivity of a given organism.

organic   See *organic* under *streambank material*.

organic debris   Material of organic origin that ranges in size from fine particulate matter to large trees.

organic particles   Particles that are of biological origin.

> **coarse particulate organic matter (CPOM)**   Living or dead organic material ranging in size from l mm (0.04 in) to 10 cm (4 in) that is often referred to as detritus.

> **dissolved organic matter (DOM) or dissolved organic carbon (DOC)**   Organic material that is smaller than 0.45 μm (i.e., passes through a 0.45 μm filter).

> **fine particulate organic matter (FPOM)** Organic material ranging in size from 0.45 μm to l mm.

> **ultrafine particulate matter**   Matter smaller than 0.45 μm.

organism   Any living thing composed of one or more cells.

orientation   Position of an object or organism relative to the direction of streamflow or current in a lake, reservoir, or ocean.

orthograde   See *orthograde* under **stratification**.

orthophotos   See *orthophotos* under **remote sensing**.

oscillation   Repeated, regular fluctuation above and below some mean value, or a single fluctuation between the maximum and minimum levels of such a periodic fluctuation. See also *oscillation* under **wave**.

outer beach   Zone of a beach that is ordinarily dry and wetted only by waves during violent storms.

outfall   Outlet of a water body, drain, culvert, or other structure.

outflow   Water flowing out of a water body, drain, sewer, or other structure.

outlet   (1) Opening or passage that allows water to flow from one place to another. (2) River or stream flowing from a water body. (3) Terminus or mouth of a stream where it flows into a larger water body such as a lake, reservoir, or sea.

outlet depth   Midline depth of principal outlet.

outlet drain   Drain that collects and transports the discharge of side drains.

outlet tower   See *outlet tower* under **control structure**.

outslope   Face of a fill or embankment that slopes downward from the highest elevation to the toe.

outsloping   Sloping a road surface to direct water away from the cut side of the road.

outwash   Material, chiefly sand or gravel, that is dislodged (i.e., washed) from a glacier by melt water.

outwash fan   Detritus mass deposited at the foot of a glacier or mouth of a gorge by free-flowing water that is heavily loaded with sediment where velocity is suddenly reduced as a result of lower lateral constrictions.

outwash plain   Flat area formed gradually by sediment carried to the site from a glacier and deposited by changes in carrying capacity of glacier meltwater.

overbank flooding   Any situation where inundation occurs when the water level of a river or stream rises above the bank.

overbank storage   Flood water from overbank flooding of a stream that remains in floodplain depressions where it is temporarily "stored" until it percolates into the stream or evaporates into the atmosphere.

overflow   See *overflow* under **mixing**.

overflow channel   Abandoned channel in a floodplain that carries water during periods of high runoff.

overhang   Organic or inorganic materials that project over or into a water body.

overhead cover   Plant foliage or overhanging material that provides protection to fish or other aquatic animals.

overlay   See *overlay* under **remote sensing**.

oversaturation   Concentrations of dissolved gases or other dissolved materials that exceed the predicted saturation level in water based on temperature and pressure.

overturn   See *overturn* under **stratification**.

oxbow   Bend or meander in a stream that becomes detached from the stream channel either from natural fluvial processes or anthropogenic disturbance.

oxbow lake   See *oxbow lake* under *lake*.

oxygen deficit   Difference between the observed oxygen concentration and the amount that would theoretically be present at 100% saturation for existing conditions of temperature and pressure.

▶ p

paleopotamon   See *paleopotamon* under *meander*.

palustrine   (1) Nontidal wetlands that are dominated by trees, shrubs, persistent or nonpersistent emergents, mosses, or lichens. (2) May also include wetlands (smaller than 8 ha) without vegetation, wetlands with water depths less than 2 m, and wetlands with salinity of less than 0.5 ppt. See also *palustrine* under *wetlands*.

parallel stream   See *parallel stream* under *stream*.

parameter   Any quantitative characteristic that describes an individual, population, or community or that describes the biological, chemical, and physical components of an ecosystem.

parapotamon   See *parapotamon* under *meander*.

parent material   Unconsolidated (more or less) weathered mineral or organic matter from which soil is developed.

partial meromixis   See *partial meromixis* under *mixing*.

particle   Individual fragment of organic or mineral material.

particle cluster   Cluster of small particles that are grouped around one or more large particles and do not move until the large particle moves.

particle size   Linear dimension, usually designated as "diameter," that characterizes the size of a particle.

particle size distribution   Frequency distribution (expressed as d_n) of the relative amounts of particles in a sample that are within a specified size range, or a cumulative frequency distribution of the relative amounts of particles that are coarser or finer than a specified size.

particulates   Any finely divided solid substance suspended in the water.

passage   An avenue or corridor for fish migration either up or down a river system.

patch   More homogeneous ecological islands that are recognizeably different from parts of an ecosystem that surround them but interact with the rest of the ecosytem.

paternoster lake   See *paternoster lake* under *lake*.

pattern   Configuration of a channel reach described in terms of its relative meander characteristics.

pavement   Surface layer of resistant material in a streambed, such as stones and rocks, exposed after finer materials have been eroded away.

pay lake   See *pay lake* under *lake*.

peak flow   See *peak flow* under *flow*.

peat   Unconsolidated, partially decomposed organic—mainly plant—material, deposited under waterlogged, oxygen-poor conditions. A layer of organic material containing plant residues that have accumulated in a very wet environment.

peat bog   See *peat bog* under *wetlands*.

peatland   General term for any land covered with a soil layer that contains a higher percentage of peat than adjoining areas.

pebble   Small (2–64 mm), gravel-sized stone with rounded edges, especially one smoothed by the action of water. Compare with other substrate sizes under *substrate size*.

pebble beach   Beach area dominated by small, gravel-sized, rounded stones.

pebble count   Method of measuring the composition of streambed material by manual collection while wading a stream.

pediment   Gradually sloping rock surface at the base of a steep slope, usually covered by thin alluvium. Also applied to a gentle sloping rock surface in front of an abrupt and receding hillslope, mountain slope, plateau, or mesa in an arid or semi-arid environment.

pelagic   (1) Open water areas of lakes, reservoirs, or seas away from the shore. (2) Refers to organisms at or near the surface in water away from the shore. Compare with *limnetic*.

pelephilic   See *pelephilic* under ***benthos***.

pellicle   Refers to a thin film or skim of material on the surface of water.

pelometabolism   Metabolism in benthic sediments, primarily bacterial metabolism in anaerobic conditions.

pelorheophilic   See *pelorheophilic* under ***benthos***.

peneplain   Low, large, nearly featureless land surface deposited by erosional processes operating over a long period of time.

peninsula   Arm of land almost completely surrounded by water.

penstock   See *penstock* under ***control structure***.

peraquic moisture regime   Soil condition where reducing conditions always occur due to the presence of groundwater at or near the soil surface.

percent fines   Percentage of fine sediments, by weight or volume, that are less than 2 mm (0.08 in) in diameter in substrate samples. See *fine sediment*.

perched groundwater   Groundwater that is separated from the main groundwater body by unsaturated, impermeable material.

perched lake   See *perched lake* under ***lake***.

perched stream   See *perched stream* under ***stream***.

percolation   Downward movement of water through soil, sand, gravel, or rock.

perennial   Stream, lake, or other water body with water present continuously during a normal water year.

perennial astatic   See *perennial astatic* under ***astatic***.

perennial flow   See *perennial flow* under ***flow***.

perennial lake   See *perennial lake* under ***lake***.

periglacial   Pertains to active valley wall processes characteristic of very cold regions such as found next to active glaciers or in alpine terrain.

periglacial landslide   See *periglacial landslide* under ***landslide***.

perilithon   Organisms (microscropic algae, protozoans, fungi, and bacteria) growing on the submerged portion of coarse rock substrates.

period   See *period* under ***wave***.

periodic drift   Drift of bottom organisms that occur regularly or at predictable time periods such as diurnal or seasonal.

periodically flooded   See *periodically flooded* under ***flooding***.

periodicity   Tendency to recur at regular intervals.

period of uninodal surface oscillation   See *period of uninodal surface oscillation* under ***wave***.

periphyton   Attached microflora growing on the bottom, or on other submerged substrates, including higher plants.

> **epilithic**   Flora growing on the surface of rock or stones.
>
> **epipelic**   Flora living on fine sediment.
>
> **epiphytic**   Flora growing on the surface of macrophytes.
>
> **epipsammic**   Flora growing on or moving through sand.

permanently flooded   Water regime where standing water covers the land surface during the entire year, except during extreme droughts. See various types of aquatic habitats in such areas under ***wetlands***. See also *permanently flooded* under ***flooding, floodplain,*** and ***water regime***.

permeability   Measure of the rate at which water can penetrate and pass through a medium such as soil or other substrate. The rate depends on the composition and degree of compaction of the substrate.

persistent emergent hydrophytes   Hydrophytes that normally remain standing, at least until the beginning of the next growing season.

pesticide   Any chemical used to control populations of organisms that are undesirable to humans. The term "pesticide" is a generic term that is applied to chemicals used to control animals. More specific terms include "herbicide" (to control plants), "insecticide" (to control insects), and "lampricide" (to control sea lampreys).

pH   Measure of the acidity and alkalinity of a solution, expressed as the negative $\log_{10}$ of the hydrogen-ion concentration on a scale of 0 (highly acidic) to 14 (highly basic). A pH of 7 is neutral.

PHABSIM   Abbreviation for "physical habitat simulation method" that is used to translate changes in streamflow to changes in quantity and quality of habitat. For a range of measured and simulated flows, the distribution of depths, velocity, substrates, and cover types across a channel are converted to an index of physical habitat as preference curves for a given species and life stage.

photic zone   See *euphotic zone.*

photogrammetry   See *photogrammetry* under *remote sensing.*

photophobic   Term applied to an organism that avoids light.

phototrophism   Movement of organisms in response to light.

phreatophyte   Deeply rooted plant that obtains its water supply from a more or less permanent subsurface zone of saturation.

physicochemical   Term applied to the physical and chemical characteristics of an ecosystem.

phytophilic   See *phytophilic* under *benthos.*

phytoplankton   See *phytoplankton* under *plankton.*

picoplankton   See *picoplankton* under *plankton.*

piedmont   Deposit that is located or formed near the base of a mountain. Also applies to a physiographic province located between mountains and a coastal plain.

pier   Structure constructed on pilings or that floats and that is used as a moorage for boats. See *pilings.*

piezometer   Small diameter, nonpumping tube, pipe, or well used to measure the elevation of water table or potentiometric surface.

pilings   Vertical columns, usually of timber, steel, or reinforced concrete, that are driven into the bottom of a water body to support a structure such as a pier or bridge.

pinnate stream   See *pinnate stream* under *stream.*

piping   Bank erosion caused by seepage of groundwater, with subsurface erosion that creates underground conduits, sometimes causing collapse of the surface.

piping depression pond   See *piping depression pond* under *pond.*

pitch   See *pitch* under *remote sensing.*

pixel   See *pixel* under *remote sensing.*

placer   Shallow deposits of gravel or similar substrate containing precious metals such as gold. The term is also applied to the site or form of mining.

placid   Term refers to surface water that is quiet with no eddies or waves and that is usually associated with very slowly moving waters.

placid flow   See *placid flow* under *flow.*

plain   Any flat or gently sloping (elevation differences of less than 150 m [500 ft]) area formed from deposition of eroded substrates at low elevations and that may be forested or bare of trees.

plane bed   Bed of fine sediments.

plankton   Small plants and animals, generally smaller than 2 mm and without strong locomotive ability, that are suspended in the water column and carried by currents or waves that may make daily or seasonal movements in the water column.

> **macroplankton**   Planktonic organisms that are larger than 500 μm.
>
> **microplankton**   Planktonic organisms that range in size from 50 to 500 μm.
>
> **nannoplankton**   Planktonic organisms that range in size from 10 to 50 μm.
>
> **phytoplankton**   Planktonic plants that are composed primarily of diatoms and algae.
>
> **picoplankton**   Planktonic organisms that are smaller than 1 μm.
>
> **ultraplankton**   Planktonic organisms that range in size from 0.5 to 10 μm.
>
> **zooplankton**   Planktonic animals that are composed primarily of protozoans and small crustaceans.

plateau Flat areas that are elevated over the surrounding terrain.

playa lake See *playa lake* under *lake*.

plesiopotamon See *plesiopotamon* under *meander*.

pleuston Organisms adapted for life in the interface between air and water.

   epipleuston Organisms living on the surface of the air–water interface.

   hypopleuston Organisms living on the underside of the air–water interface.

plug A piece of material or an accumulation of material that prevents the movement of water or other fluids.

plume (1) Flow of dissolved or suspended material into a larger water body. (2) Mass of water discharged by a river, outfall, or some other source into a water body that is not completely mixed and retains measurably different characteristics from the rest of a water body.

plunge lake See *plunge lake* under *lake*.

plunge pool See *plunge pool* under *pool, scour pool* under the main heading *channel unit*.

plunging breaker See *plunging breaker* under *wave*.

pluvial Refers to rain or the action of falling rain.

pluvial lake See *pluvial lake* under *lake*.

pocket water See *pocket water* under *pool, scour pool* under the main heading *channel unit*.

pocosin See *pocosin* under *wetlands*.

poikilohaline See *poikilohaline* under *salinity*.

poincare wave See *poincare wave* under *wave*.

point Peninsula or land that projects from the shore into a water body. A stationary location used for reference such as the single source of a pollutant.

point bar See *point bar* under *bar*.

point of diversion Location where water is diverted for some use.

point of inflection See *point of inflection* under *meander*.

point source Material, usually pollutants, flowing into a water body from a single well-defined source such as a pipe or ditch.

pollute To contaminate land, water, air, plants, animals, or microorganisms with substances considered objectionable or harmful to the health of living organisms.

pollution Presence of matter or energy, usually of human origin, whose nature, location, or quantity, produces undesired environmental effects on natural systems.

polygon See *polygon* under *remote sensing*.

polyhaline See *polyhaline* under *salinity*.

polymictic See *polymictic* under *mixing*.

polysaline See *polysaline* under *salinity*.

polysaprobic zone See *polysaprobic zone* under *saprobien system*.

pond Natural or artificial body of standing water that is typically smaller than a lake (less than 8 ha [20 acre]), characterized by a high ratio of littoral zone relative to open water.

   aerated pond or lagoon Natural or artificial basin where mechanical equipment is used to increase the supply of oxygen to decompose organic waste, increase aquaculture production, or improve aesthetics.

   aestival pond Pond that exists only in summer.

   anchialine pool Mostly small, irregular water exposures in barren lava with a surface level at the marine water table so the water is mixohaline from dilution by groundwater and salinity is usually 1–10‰.

   beaver pond Pond containing water impounded behind a dam built by beaver.

   borrowpit pond Pond formed by the accumulation of water in an area excavated by mining for sand, gravel, or boulders used in construction.

   charco Small pond formed by a soil dam that is usually round with a basin shaped like an inverted cone that is filled by groundwater.

Often found in the desert of southwestern United States.

cooling pond   Pond with water used to cool equipment in a power plant or other industrial facility.

dug pond   Pond formed by excavation, without a dam, and supplied with water from runoff or seepage.

dune pond   Pond in a basin formed from the blockage of a stream mouth by sand dunes that move along a shoreline of an ocean.

evaporation pond   Shallow ponds filled with water that are allowed to evaporate to recover suspended or dissolved materials such as salts or other minerals.

farm pond   Pond created for agricultural purposes (i.e., irrigation or water for livestock), culture of commercial fishes, or for recreation (including sportfishing, swimming, and boating).

fault sag pond   Pond in a small depression along an active or recent geologic fault that is supplied by groundwater.

holding pond   Pond or reservoir constructed for settling and storage of sediments, for aerating or aging water for a fish hatchery, or for storing wastes or polluted runoff.

lagoon   See *lagoon*.

log pond   Pond used for storing logs, generally attached to saw or veneer mills. See *mill pond*.

mill pond   Impoundment created by damming a stream to produce a head of water for operating a mill or for storage of logs. Synonymous with *log pond*.

off-channel pond   Pond that is not part of the active channel but is supplied with water from overbank flooding or through a connection with the main stream by a short channel. These ponds are generally located on flood terraces and are called wall-based channel ponds when located near the base of a valley wall.

piping depression pond   Pond that forms in a small depression resulting from subsurface piping.

quarry   Pond formed in the depression created by excavation of rock or coal (as well as clay in

some instances) and generally supplied with groundwater.

sag pond   Small body of water occupying a depression or sag formed by active or recent geologic fault movements.

salt chuck pond   A log pond in seawater. Compare with *salt chuck*.

sedimentation pond   Impoundment created to trap suspended sediments. See *settling pond* and *stilling pond* under **pond**.

settling pond   Impoundment used to precipitate materials that accumulate on the bottom and are removed periodically. See *sedimentation pond* and *stilling pond* under **pond**.

stilling pond   Deep depression constructed in a streambed on the outwash fan that is used to catch detritus and sediments. See *sedimentation pond* and *settling pond* under **pond**.

tank   Artificial pond to hold water for livestock, wildlife (sometimes including fish) and other uses.

vernal pond   Small (usually less than 1 ha), temporary pond that forms from melting snow and rainfall in late winter or early spring.

wall-based pond   See *off-channel pond* under **pond**.

pondage   Term applied to storage capacity or water that is held for later release above the dam of a hydroelectric plant to equalize daily or weekly fluctuations of streamflow or to permit irregular releases of water through turbines to accommodate the demand for electricity.

ponded   Water that is impounded from anthropogenic or natural blockage or obstruction. Also referred to as ponding.

pool   Small depression with standing water such as found in a marsh or on a floodplain. Also see *pool* under *slow water* under the main heading **channel unit**.

pool digger   Rock or log structure designed to scour a plunge pool on its downstream side or a lateral scour pool.

pooled channel   See *pooled channel* under **stream**.

pool feature   Condition or object that causes the formation of a pool including logs, trees, roots, stumps, brush, debris, channel meanders, sediment deposition, beaver dams, culverts, bridges, or other artificial or natural structures.

pool margin   Outer edge of a pool as identified by bed topography.

pool quality   Estimate of the ability of a pool to support target fish species, based on measurements such as length, width, depth, velocity, and cover.

pool : riffle ratio   See *pool : riffle ratio* under *dimensions*.

poorly drained   Condition where water is removed from the soil so slowly that the soil is saturated periodically during the growing season and remains wet for long periods (e.g., more than 7 d).

pore space   Unoccupied interstices in the substrate.

porosity   (1) Existence of interstices or "pores" in soil or rock, and the ratio of the volume of pores to the total volume of solids plus voids. (2) Also refers to the ease or speed that water can move into or through the substrate. See also *porosity* under *groundwater*.

positive estuary   Coastal indentures where there is a measurable dilution of seawater by land drainage so that freshwater inflow plus precipitation is greater than evaporation.

positive heterograde   See *positive heterograde* under *stratification*.

potable   Water that is suitable or safe for drinking according to established health standards.

potamodromous   Life-cycle strategy of a fish that includes migrations, spawning, and feeding entirely in freshwater. Compare with *anadromous, catadromous, diadromous, oceanadromous*.

potamology   Study of the biological, chemical, geological, and physical aspects of rivers.

potamon   Portion of a stream that includes the thalweg (or deepest part of the channel) and is nearly always defined as lotic. Also applied to that portion of a stream that contains water even if discharge becomes intermittent.

  epipotamon   Upper layer of the potamon region.

  hypopotamon   Bottom layer of the potamon region.

  metapotamon   Intermediate layer of the potamon region.

potamon plankton   Plankton living in freshwater lotic habitats.

potential energy   See *potential energy* under *energy*.

pothole lake   See *pothole lake* under *lake*.

power pool   See *operating pool*.

prairie pothole   Ponds, pools, lakes, and wetlands found in depressions (potholes) that were formed by glacial activity. A local term used in the Great Plains of central United States and Canada. See *pothole lake* under *lake*.

precipitate   Solid that settles from water through physical or chemical changes.

precipitation   Water, hail, sleet, and snow that falls to earth from its origin as atmospheric moisture.

predators   See *predators* under *macroinvertebrate*.

preservation   Protection and maintenance of intact and functional natural areas and ecosystems. Compare with *restoration*.

pressure   A force per unit of area.

pressure head   See *pressure head* under *dimensions*.

pristine   Term used to describe a natural location or habitat unaffected by anthropogenic disturbances.

problem area wetland   See *problem area wetland* under *wetlands*.

process water   Water used in manufacturing or processing including the production or use of any raw material, intermediate product, finished product, by-product, or waste product. See *effluent*.

production   (1) Process of producing organic material. (2) Increase in biomass by individuals, species, or communities with time (e.g., the total amount of fish tissue produced by a population of fish within a specified period of time).

  gross primary production   Total rate of photosynthesis including the organic matter used up in respiration during the measurement period.

net primary production   Rate of storage of organic matter in plant tissues in excess of the respiratory use by the plants during the measurement period.

secondary production   Total energy storage at the consumer and decomposer trophic levels. Consumers and decomposers utilize food materials that have already been produced and convert this matter in different tissues with energy loss to respiration. Efficiency of conversion in secondary production decreases with trophic levels.

productivity   (1) Rate of formation of new tissue or energy use by one or more organisms. (2) Capacity or ability of an environmental unit to produce organic material. (3) Recruitment ability of a population from natural reproduction.

profile   Graphical or other representation of shape or relationship. Compare with *stream profile*.

profundal   Deep, bottom-water area beyond the depth of effective light penetration including all of the lake or sea floor beneath the upper margin of the hypolimnion.

progressive wave   See *progressive wave* under *wave*.

prolonged speed   See *prolonged speed* under *swimming speed*.

promontory   High point of land or rock overlooking or projecting into a water body.

protected area   Area administratively set aside as a buffered or structured area that is shielded from damage from anthropogenic disturbances.

psammon   Refers to the beach zone along an ocean or an organism growing in or moving through sand on the beach.

psammorheophilic   See *psammorheophilic* under *benthos*.

public lake   See *public lake* under *lake*.

public water   Water that is navigable by federal test or court decree, legally accessible over public lands, or impressed with prescriptive rights vested in the public.

public water system   System providing piped water for public consumption.

puddle   Small (several inches to several feet in its greatest dimension), shallow (usually a few inches in depth) pool of water that is ephemeral and often dirty or muddy.

pump chance   Water body, usually small, that is accessible to equipment collecting water for fire control.

pumped storage reservoir   Reservoir used to store water for use during peak periods of electricity production. During periods of low demand for electricity, water is pumped into the storage reservoir for later release through turbines to generate electricity during periods of peak demand.

pycnocline   See *pycnocline* under *stratification*.

▶ q

quaking bog   See *quaking bog* under *wetlands*.

quarry   See *quarry* under *pond*.

quick sand   Soft or loose sand, sometimes of measurable depth, saturated with spring or stream water that yields when weight is placed on it.

▶ r

radial   See *radial* under *stream*.

radial gate   See *radial gate* under *gate* and main heading of *control structure*.

radio telemetry   See *radio telemetry* under *remote sensing*.

radius of curvature   See *radius of curvature* under *dimensions* as well as under *meander*.

raft   (1) Collection of timbers or bamboo, tied together or enclosed within a boom, for transport by floating. (2) Also, a large number of closely spaced timbers that can free float on water.

rain-on-snow event   Event that occurs in late winter, spring, and early summer when snowpacks are partially or completely melted during rainstorms causing flooding.

rain shadow   Area of reduced precipitation on the leeward side of mountains that results from the interception of storms by the mountains.

raised bog   See *raised bog* under *wetlands*.

rapids   See *rapids* under *fast water—turbulent* with the main heading of *channel unit*.

raster   See *raster* under *remote sensing*.

raster data   See *raster data* under *remote sensing*.

ravine   Narrow, steep-sided valley that is commonly eroded by running water.

raw water   Untreated surface or groundwater that is available for use but may or may not be potable.

reach   (1) Any specified length of stream. (2) Relatively homogeneous stretch of a stream having a repetitious sequence of physical characteristics and habitat types. (3) Length of channel where a single gage affords a satisfactory measure of the stage and discharge. (4) Portion of a stream that extends downstream from the confluence of two streams or rivers to the next encountered confluence. (5) According to the Environmental Protection Agency, that portion of a river reach extending downstream from the confluence of two rivers (or from the uppermost end of a river) to the next encountered confluence.

  archival reach   Reach whose boundaries and properties have been modified by natural or other events since an original survey was conducted. Original data is compared with new data to evelute changes.

  critical reach   Stream segment that is essential for development and survival of a particular aquatic organism, or a particular life stage of an aquatic organism.

  representative reach   Stream segment that represents a larger segment of the stream with respect to area, depth, discharge, slope, or other physiochemical or biotic characteristics.

  specific reach   Stream segment that is uniform with respect to selected habitat characteristics or elements (discharge, depth, area, slope, population of hydraulic units), species composition, water quality, and condition of bank cover.

reaeration   Supply of oxygen to oxygen-depleted water.

rearing habitat   Areas in a body of water where larval and juvenile fish find food and shelter to live and grow. Also referred to as nursery habitat.

reattachment bar   See *reattachment bar* under *bar*.

receiving site   Water collection sites where water is collected by subsurface flow from higher elevations and precipitation.

receiving waters   Any body of water into which untreated or treated wastes, or polluted waters, are discharged.

recessional moraine   See *recessional moraine* under *moraine*.

recharge   Process by which water is added to an aquifer.

recharge area   Area where water infiltrates into the ground and joins the aquifer through hydraulic head.

recharge zone   Area through which water is added to an aquifer.

reclamation   (1) Most recently defined as any action that results in a stable, self-sustaining ecosystem that may or may not include introduced species. (2) Traditionally defined as the process of adapting natural resources to serve a utilitarian human purpose. Historically this term included the conversion of riparian or wetland ecosystems to agricultural, industrial, or urban uses.

recovery   Ability of a disturbed system to reestablish habitats and plant or animal communities that were present prior to the disturbance.

recreational pool   The "normal" surface elevation of a reservoir with fluctuating water levels that is generally a stable level established by the water controlling agency to maximize recreational uses during spring, summer, and fall (i.e., open water period). Compare with *summer pool*.

rectangular drainage   See *rectangular stream* under *stream*.

rectangular stream   See *rectangular stream* under *stream*.

rectification   See *rectification* under *remote sensing*.

recurrence interval   Expected or observed time intervals between hydrological events of a particular magnitude described by stochastic or probabilistic modes (log-log plots). The average interval of time within which a given event, such as a flood, will be equalled or exceeded one time.

redd   Nest excavated in the substrate by fish for spawning where fertilized eggs are deposited and develop until the eggs hatch and larvae emerge from the substrate.

reducers   Organisms, usually bacteria or fungi, that break down complex organic materials into simpler compounds.

reef   Ridge of rocks, sand, soil or coral projecting from the bottom to or near the surface of the water.

reference wetland   See *reference wetland* under *wetlands*.

reflectance   See *reflectance* under *remote sensing*.

reflected solar radiation   See *reflected solar radiation* under *solar radiation*.

refracted solar radiation   See *refracted solar radiation* under *solar radiation*.

refraction   See *refraction* under *wave*.

refugium   (1) Habitats that support sustainable populations of organisms that are limited to fragments of their previous historic and geographic range. (2) Habitats that sustain organisms during periods when ecological conditions are not suitable elsewhere. For example, trout in alpine areas use the deeper pools in a stream during winter or fish use a lake with high dissolved oxygen levels to escape adjacent hypoxic swamps and marshes. (3) Waters where threatened or endangered fishes are placed for safe-keeping or where a portion of the population is maintained to prevent extinction.

regime   (1) Seasonal pattern of streamflow during a year. (2) Balance or equilibrium of erosion and deposition in a channel with time so that the stream channel maintains its overall characteristics.

regimen   Characteristics of a stream with respect to velocity and volume changes in a channel capacity to transport sediment and the amount of sediment transported with time.

regolith   Unconsolidated mantle of weathered rock, soil, and surficial materials overlying solid rock.

regular meander   See *regular meander* under *meander*.

regular meander channel   See *regular meander channel* under *channel pattern*.

regulated flow   See *regulated flow* under *flow*.

regulated zone   Area in a reservoir between conservation pool and flood control pool.

regulation   Control of the volume and timing of streamflow at a specific location.

rehabilitation   (1) Action taken to return a landform, vegetation, or water body to as near its original condition as practical. (2) Term implies making land and water resources useful again (primarily for humans) after natural or anthropogenic disturbances. This term differs from restoration that implies a return to predisturbance conditions and functions in natural aquatic or terrestrial systems. See *restoration*.

rejuvenated river   See *rejuvenated river* under *river*.

relative depth   See *relative depth* under *dimensions*.

relative thermal resistance   Ratio of the density difference between water at the top and bottom of a water column with a definite thickness, to the density difference between water at 5°C and 4°C. "Thermal resistance" also refers to mixing, or the amount of work done by meteorological or anthropogenic events to mix a water column.

relict   Remnant of a biotic community or population that was formerly widespread.

relict lake   See *relict lake* under *lake*.

relief   (1) Change in elevation of a land surface between two points. (2) Configuration of the earth's surface including such features as hills and valleys.

relief ratio   See *relief ratio* under *dimensions*.

remote sensing   (1) Acquisition of information from a distance, generally by transmissions that involve electromagnetic energy and sometimes by gravity and sound. (2) Measurement or acquisition of data on an object by satellite, aerial photography, and radar that are some distance from the object.

   aerial photography   Images on film or in digital format taken above the surface of a planet.

attribute Nongraphic information associated with a point, line, or polygon.

AVHRR Advanced very high resolution (1.1 × 1.1 km or 4 × 4 km) radiometer data with small-scale imagery produced by a NOAA polar orbiting satellite.

azimuth Principal plane in a clockwise angle on a tilted photograph.

band Set of values for a specific portion of the electromagnetic spectrum of reflected light, emitted heat, or some other user-defined information, created by combining or enhancing the original bands.

band ratio Method in which ratios of different spectral bands from the same image or from two registered images are used to reduce certain effects such as topography or to enhance subtle differences of certain features.

cartesian Coordinate system in which data are organized on a grid and points on the ground are referenced by their x- and y-coordinates.

cell Pixel or grid cell with a 1° × 1° area of coverage.

cell size Area represented by one pixel, measured in map units.

class Set of pixels in a GIS file that represents areas that share some condition.

clump Contiguous group of pixels in one class that is also called a raster region.

cluster Natural groupings of pixels when plotted in a spectral space.

conjugate points (conjugate principal points) Positions on an aerial photograph that are principal points on adjacent photos of the same flight line. Conjugate points are used to justify or align photographs for creation of a larger photographic mosaic.

digital classification Process for using algorithms to group pixels with similar spectral signatures.

digital enhancement Manipulating digital information to increase or improve features of interest for interpretation.

digital terrain model Analysis of pixel or topographic information to produce a three-dimensional representation of the landform. Also called digital elevation model.

digitizing Process that converts nondigital data into numerical data that is usually stored in a computer.

false color Use of one color to represent a characteristic or feature or a color substituted for the true color on an image.

geometric registration Process of aligning data resolution scales so that information can be visually or digitally superimposed.

ground cover (1) Vegetation and litter on or slightly above the ground surface. (2) The percentage of area bearing such cover.

ground truthing Field verification of data gathered away from the site by remote sensing or by some other method.

image Picture or representation of an object or scene on paper or computer display screen. Remotely sensed images are digital representations of the earth.

landsat Satellite system that provides imagery used for remote sensing inventory and analysis. Also refers to a series of earth-orbiting satellites gathering multispectral scanner or thematic mapper imagery.

landsat multispectral scanner (MSS) Satellite-borne sensor capable of recording reflectance energy from the surface of the earth in four wavelength bands for a 180 × 180 km scene.

landsat thematic mapper (TM) Satellite sensor capable of recording reflected and emitted energy from the surface of the earth in seven bands or divisions of the visible and infrared spectrum.

maximum likelihood Classification decision rule based on the probability that a pixel belongs to a particular class. The basic equation assumes that the probabilities are equal for all classes, and that the input bands have normal distribution.

merge The process of combining information from one or more sources or the restructuring of an existing database to create a new database that retains the original data.

mosaic (1) Pattern of vegetation across a landscape. (2) Composite image created by joining smaller images, usually aerial photographs, into a single composite.

multispectral classification Process of sorting pixels into a finite number of individual classes or categories of data that are based on data files in multiple bands.

multispectral imagery   Satellite imagery with data recorded in two or more bands.

multispectral scanner (MSS)   Landsat sensor system that generates spectral data from reflected light in the visible light spectrum.

nadir   Point where a vertical line from the center of the camera lens intersects the plane of the photograph.

orthophotos   Images based on aerial photographs that are true to scale and free of distortion. Orthophotos resemble aerial photographs but are really accurate maps.

overlay   Process in which data from different themes or plots are placed over a base map or in a series to show spatial interactions.

photogrammetry   Gathering of information on physical objects and the environment by recording and interpreting images and phenomena.

pitch   Rotation of a camera around the y- or exterior x-axis.

pixel   Abbreviation for "picture element" that is the smallest division of an image.

polygon   Closed figure usually with three or more sides. Also refers to a set of closed lines defining an area.

radio telemetry   Use of transmitters attached to an animal to send signals to a remote receiver that is used to track the animal.

raster   Pattern of scanning lines that cover an area where images are projected.

raster data   Data that are organized in a grid of columns and rows and usually represent a planar graph or geographic area.

rectification   Transformation of an image to a horizontal plane to correct tilt and to convert to a desired scale.

reflectance   Measure of the ability of a surface to reflect energy as a ratio of reflected and incident light. Reflectance is influenced by the nature of the reflected surface and the pattern of light.

remote survey   Measurement or acquisition of information by a recording device that is not in physical contact with the object under study. More precisely, recording of environmental images using electromagnetic radiation sensors and their interpretation.

resolution   (1) Spatial: Ability to reproduce an isolated object or to separate closely spaced objects or lines that are usually measured in lines per millimeter. (2) Temporal: How often a sensor records imagery of a specific geographcal area. (3) Spectral: The number and dimension of wavelength intervals in the electromagnetic spectrum recorded by the sensor. (4) Radiometric: Sensitivity of a detector to differences in signal strength.

satellite imagery   Passive images of natural radiation detected in visual or infrared wavelengths.

scan line   Strip of land within the view of a remote sensor as it passes over a surface.

scanning   Transfer of analog data, such as photographs, maps, or other viewable images into a digital (raster) format.

sensor   Device that gathers energy, converts it to a digital value, and presents it in a form suitable for obtaining information about the environment.

signature   Set of statistics that define a training sample or cluster that is used in a classification process. Each signature corresponds to a GIS class that is created with a classification decision rule.

SLAR   Abbreviation for "Sideways Looking Airborne Radar." A form of remote sensing where an aircraft sends out and receives long wavelength radiation, and interference in the return pattern is analyzed for physical features of the area surveyed. An antenna is fixed below an aircraft and pointed to one side to transmit and receive the radar signal.

sonar   Method using echolocation to detect and locate objects, including living organisms, below the surface of water.

spectral signature (spectral reflectance curve)   Characteristic wavelength patterns associated with vegetation, structures, water, or other features.

SPOT   Series of earth-orbiting satellites operated by the Centre National d'Etude Spatiales of France.

supervised classification Computer-implemented classification based on pattern recognition of assigned class signatures.

supervised training Any method of generating signatures for classification in which the analyst is directly involved in a pattern recognition process. Supervised training usually requires an analyst to select training samples from the data that represent patterns to be classified.

swath width Total width of the area on the ground covered by the scanner in a satellite system.

synthetic aperture radar (SAR) Use of a side-looking, fixed antenna sensor to create a synthetic aperture. The sensor transmits and receives as it is moving, and the signals that are received during a time interval are combined to create an image. SAR sensors are mounted on satellites and the NASA space shuttle.

thematic data Raster data that are qualitative and categorical. Thematic layers often contain classes of related information, such as land cover, soil type, slope, and hydrology that can be displayed on maps illustrating the class characteristics.

thematic mapper Advanced satellite sensor system in Landsat 4 and 5 that incorporates radiometric and geographic design improvements relative to the older MSS system.

theme Data set used for mapping information on a particular subject. Individual theme information can be displayed and related to other themes.

training Process of defining the criteria by which patterns in image data are recognized for classification.

unsupervised classification Computer-automated method of pattern recognition where some parameters are specified by the user to define statistical patterns that are inherent in the data.

vector A line in space characterized by direction and magnitude. In GIS systems, vectors are used to create a file of points that can be connected from point to point to create line segments.

vector data Data that represent physical elements such as points, lines, and polygons.

In a remote sensing database, only verticals of vector data are stored (rather than every point that makes up the element).

vector format A GIS database file where information is used to code lines and polygons to express size, direction, and degree of connection between data points.

remote survey See *remote survey* under **remote sensing**.

repose bank See *repose bank* under **streambank**.

representative reach See *representative reach* under **reach**.

reregulating reservoir Reservoir for reducing diurnal fluctuations in volume from the operation of an upstream reservoir for power production.

reservoir (1) Generally, natural or artificial impoundment where water is collected, stored, regulated, and released for human use. (2) An underground porous, permeable substrate that contains accumulated water. See *reservoir* under **lake** and **small impoundment**.

reservoir river See *reservoir river* under **river**.

residence time Amount of time that some material, such as large woody debris, pesticide, or sedimentary material, remains in one location.

resident fish Fish species that remain in one water body (i.e., nonmigratory species).

resident species Organisms normally found in a single habitat, ecosystem, or area.

residual depression storage Depression storage that exists at the end of a period of heavy rainfall.

residual depth Term corresponding to a minimum streamflow that just barely flows through pools that is calculated by subtracting water depth at a riffle crest from water depth in the upstream pool.

residual detention storage Detention storage existing at the end of a period of heavy rainfall.

residual pool See *residual pool* under *slow water, pool, scour pool* under the main heading **channel unit**.

residual volume of fine sediment Fraction of a

scour pool volume (V^*) occupied by fine sediment.

$$V^* = \frac{V_f}{(V_f + V_r)} \; ;$$

V_f = fine sediment volume;
V_r = residual pool volume.

resilience Capacity of species or ecosystems to recover after a natural disturbance or anthropogenic perturbation.

resistance Capacity of an ecosystem to maintain natural function and structure after a natural disturbance or anthropogenic perturbation.

response segment Reach or segment of a stream channel where localized inputs of wood, water, energy, and sediments causes changes in form to that reach of the stream channel.

resolution See *resolution* under **remote sensing**.

restoration (1) Reestablishment of predisturbance riparian or stream functions and related biological, chemical, and physical processes in an ecosystem. (2) Actions taken to return a habitat, an ecosystem, or a community to its original condition after damage resulting from a natural disturbance or an anthropogenic perturbation. (3) Sometimes used to describe reestablishment of fish stocks or populations that were eliminated or reduced from anthropogenic actions. See *rehabilitation*.

restrictive layer Soil layer that restricts the movement of water because of its density or composition.

resurgent water Water that resurfaces or reappears.

retard See *fence barrier* under **habitat enhancements**.

retarding reservoir See *detention reservoir*.

retention Portion of the gross storm rainfall that is intercepted, stored, or delayed, and thus fails to reach a concentration point by either surface or subsurface routes during the time period under consideration.

retention time Length of time that water is stored within a drainage system or water body.

return flow See *return flow* under *flow*.

revetment A facing or structure made of hard material such as boulders or logs along a streambank or shoreline that reduces erosion. See *riprap* under **habitat enhancements**.

Reynolds number (R_e) Dimensionless value expressing the ratio of inertial to viscous forces acting on a fluid or a particle in the fluid:

$$R_e = \frac{VD}{v} \; ;$$

V = velocity of the fluid or particle (m/s);
D = a relevant dimension (pipe diameter, particle length, etc.) (m);
v = kinetic viscosity (10^{-6} m^2/s).

rheocrenes Perennial seeps and springs that flow only a short distance over a rock surface or in indistinct channels.

rheophilus Current-loving organisms.

rheophyte See *rheophyte* under **macrophyte**.

rhithron Reach of stream that extends from the headwaters downstream to where the mean monthly summer temperature reaches 20°C, dissolved oxygen levels are always high, flow is fast and turbulent, and the bed is composed of rocks or gravel with occasional sandy or silty patches. The rhithron is subdivided into three zones covering a range of water courses.

epirhithron Upper reaches of the rhithron region that are characterized by rapids, waterfalls, and cascades.

hyporhithron Lower reaches of the rhithron region that are characterized by an increase in backwaters with mud and debris bottoms.

metarhithron Middle reaches of the rhithron region that are characterized by a more gentle gradient and higher percentage of pools.

ribbon falls See *ribbon falls* under *fast water—turbulent*, *falls* under the main heading **channel unit**.

riffle crest Shallowest continuous line (usually not straight) across the channel close to where a water surface becomes continuously riffled.

riffles See *riffles* under *fast water—turbulent* under the main heading **channel units**.

riffle stability index   An index to determine the size class percentage of riffle material moved during channel forming flows. Determined by comparing the largest, commonly occurring size of particles moved by the force of a frequent flood event to the cumulative particle size distribution of bed materials in a riffle

rift lake   See *rift lake* under **lake**.

rift valley   Long, narrow valley resulting from subsidence (i.e., settling) of strata between more or less parallel geologic faults.

rill   One of the first and smallest channels formed by surface runoff.

rill erosion   Mild water erosion caused by overland flow producing very small and numerous channels. See also *rill erosion* under **erosion**.

rilling   Removal of soil by water from very small but well-defined, visible channels or streamlets where there is substantial overland flow.

riparian area   (1) Of, pertaining to, situated or dwelling on the margin of a river or other water body. (2) Also applies to banks on water bodies where sufficient soil moisture supports the growth of mesic vegetation that requires a moderate amount of moisture. Also referred to as riparian zone, riparian management area, or riparian habitat.

riparian ecosystem   Ecosystem located between ecocline of aquatic and terrestrial environments. See *ecocline*.

riparian rights   Entitlement to water on or bordering a landowner's property including the right to prevent diversion of water upstream.

riparian vegetation   Vegetation growing on or near the banks of a stream or other water body that is more dependent on water than vegetation that is found further upslope.

riparian vegetation erosion control rating   System for ranking the relative effectiveness of riparian vegetation to control bank erosion.

ripping   The process of breaking up or loosening compacted soil to improve aeration and assure development of root systems from seeds or planted seedlings.

ripple   See *ripple* under **wave**.

riprap   See *riprap* under **habitat enhancements**. Compare with *revetment* under **habitat enhancements**.

rithron   See *rhithron*.

river   Large, natural or human-modified stream that flows in a defined course or channel, or a series of diverging and converging channels.

desert river   River in an arid area that is characterized by flash floods and no tributaries. Desert rivers increase in alkalinity and conductivity as they flow downstream and may terminate in salt marshes or lakes.

flood river   River with extremes of annual fluctuation in streamflow.

mature river   River system where erosion and deposition are in balance.

old river   River where depositional processes dominate.

rejuvenated river   An old or mature river where gradient changes result in a temporary reversal of normal succession processes.

reservoir river   River with an extensive area of lakes, swamps, and floodplain depressions that stores or holds water.

sandbank river   River that conveys floodwaters but frequently ceases to flow or even dry out seasonally.

savanna river   Floodbank or sandbank river characterized by high silt loads and low pH and conductivity.

tropical river   River in the tropics that functions similar to a reservoir river and is characterized by black water with low pH, low conductivity, low silt load, and high humus load.

tundra river   River in an arctic or subarctic region with streamflows that fluctuate with the freezing cycle.

young river   Generally used in reference to the headwaters where erosional processes are most active.

riverbank   Elevated edges of a channel that control lateral movement of water.

river channel   Natural or artificial open conduits that continuously or periodically contain moving water. Also applied to a connection between two water bodies.

river continuum   Ecological succession that occurs from the headwaters to the mouth in a river and that is associated with an increase in nutrients and organic matter.

riverine (riverain)   (1) Habitats that are formed by or associated with a river or stream. (2) Wetlands and deeper water habitats within a channel that are influenced strongly by the energy of flowing water. (3) Also applied to vegetation growing in a floodplain, in close proximity to water courses with flowing water, or on islands in a river.

riverine wetland   See *riverine wetland* under *wetlands*.

rivulet   Refers to a small stream.

rock   Mass of stone of any size, consolidated or unconsolidated, of various mineral composition.

rock avalanche   See *rock avalanche* under *landslide*.

rock creep   See *rock creep* under *landslide*.

rock fall   See *rock fall* under *landslide*.

rock-fill dam   (1) Dam composed of large, broken, and loosely placed rocks that allows water to percolate and continue to flow downstream. (2) Dam with an impervious core of composted rock and soil with large rocks (i.e., riprap) on the upstream face or surface.

rock glacier   See *rock glacier* under *glacier*.

rock slide   See *rock slide* under *landslide*.

roller dam   See *roller dam* under *habitat enhancements*.

roller gate   See *roller gate* under *gate* and main heading *control structure*.

rolling flow   See *rolling flow* under *flow*.

root wad   See *root wad* under *large organic debris*.

rotational failure   See *rotational failure* under *landslide*.

roughness   Pertains to the irregularity of a substrate surface.

roughness coefficient   See *Manning's n*.

roughness element   Any object or structure (e.g., bedrock outcrops, large woody debris, and boulders) that obstructs streamflow in a channel and influences the pattern of bed load transport and deposition in a stream reach.

rubble   Stream substrate particles between 128 and 256 mm (5–10 in) in diameter. Compare with other substrate sizes under *substrate size*.

run   See *run* under *fast water—nonturbulent* under the main heading *channel unit*.

runoff   (1) Natural drainage of water away from an area. (2) Precipitation that flows overland before entering a defined stream channel. (3) Total discharge of stream within a specified time from a specific area that includes both surface and subsurface discharge and is generally measured in cubic feet (cubic meters) or acre feet (hectare meters) of water.

runoff curve   Graphic estimator of runoff potential in a drainage basin based on precipitation, soils, vegetation, and land use.

run-of-the-river flow   Flow through a dam that is minimally regulated by the dam and approximates the flow that would occur in the absence of a dam.

run-of-the-river reservoir   Narrow reservoir that is held to the width of the natural river channel in which short-term water input approximates equal short-term outflow.

▶ S

saddle   Narrow, submerged isthmus of land surrounded by deeper water which may connect a point and an island, hummock, or another land mass.

sag pond   See *sag pond* under *pond*.

salinas   Inland desert basins that are light-colored from salt.

saline   (1) Soil or water containing sufficient soluble salts to interfere with the growth of most plants. (2) See *saline* under *salinity*.

saline marsh   See *saline marsh* under *wetlands*.

saline lake   See *saline lake* under *lake*.

salinity   Relative concentration of salts, mainly sodium chloride, in a given water, expressed as the weight per volume or weight per weight. The terms "haline" and "saline" are often used interchangeably, but differ based on the origin of the salts. Haline refers to ocean-derived salts and saline refers to land-derived salts.

euryhaline (1) Waters with a salinity between 30.1 and 40‰ (parts per thousand) from ocean-derived salts. (2) Organisms that are able to live in waters with a wide range of ocean-derived salts.

eurysaline (1) Waters with a salinity between 30.1 and 40‰ from land-derived salts. (2) Organisms that are able to live in waters with a wide range of land-derived salts.

freshwater Water with salinity of less than 0.5‰ dissolved salts.

haline Refers to saline conditions from ocean-derived salts.

halocline Well-defined vertical salinity cleavage or boundary in a water body.

homoiohaline Refers to saline conditions in oceans that are either stable or with narrow fluctuations.

hydrohaline Waters with a salinity greater than 40‰ from ocean-derived salts. Also referred to as hyperhaline.

hydrosaline Waters with salinity greater than 40‰ from land-derived salts. Also referred to as hypersaline.

mesohaline (1) Waters with salinity between 5.1 and 18‰ from ocean-derived salts. Also referred to as metahaline. (2) Saltwater organisms that are able to live in waters with medium salinities.

mesosaline (1) Waters with salinity between 5.1 and 18‰ from land-derived salts. Also referred to as metasaline. (2) Organisms that are able to live in waters with a medium range of salinities from land-derived salts.

mixohaline Waters with salinity between 0.5 and 30‰ from ocean-derived salts.

mixosaline Water with salinity between 0.5 and 30‰ from land-derived salts.

oligohaline (1) Waters with salinity between 0.5 and 5.0‰ from ocean-derived salts. (2) Saltwater organisms that are able to live in waters with low salinities.

oligosaline (1) Waters with salinity between 0.5 and 5.0‰ from land-derived salts. (2) Organisms that are able to live in waters with a low range of salinities from land-derived salts.

poikilohaline Salt concentrations that fluctuate widely.

polyhaline Waters with salinity between 18.1 and 30‰ from ocean-derived salts.

polysaline Waters with salinity between 18.1 and 30‰ from land-derived salts.

saline Waters with salinity that is greater than 30‰.

seawater Waters with salinity of about 35‰ dissolved salts.

stenohaline Organisms that are able to live in waters with a narrow range of ocean-derived salts.

stenosaline Organisms that are able to live in waters with a narrow range of land-derived salts.

saltation erosion See *saltation erosion* under *erosion*.

salt chuck General term for the estuarine areas at the mouth of rivers. Compare with *salt chuck* under *pond*.

saltern lake See Saltern Lake under *lake*.

salt flat Land area with little or no elevational changes and with a surface layer of salts that remains after prolonged flooding and desiccation.

salt karst lake See *salt karst lake* under *lake*.

salt marsh See *salt marsh* under *wetlands*.

salt water intrusion Invasion of saltwater into fresh surface or groundwater systems, usually as a result of freshwater depletion that provides access for saltwater.

sand Substrate particles between 0.062 and 2 mm (0.00003–0.01 in) in diameter. Compare with other substrate sizes under *substrate size*.

sandbank river See *sandbank river* under *river*.

sand dune See *sand wave*.

sand splay   Deposits of flood debris that are usually composed of coarse organic matter and sand particles in the form of splays or scattered debris.

sand wave   Series of generally sinusoidal waves that form on the sandy bottom of a river from the interaction of flowing water and the substrate when the Froude number is close to or greater than one. Sand waves are often transitory, vary in height, and may migrate along the river bottom. Also referred to as an *antidune* or *sand dune*.

saprobic   Term applied to living on dead or decaying organic matter.

saprobicity   Sum of all metabolic processes that can be measured either by the dynamics of metabolism or analysis of community structure.

saprobien system   System of classifying organisms according to their response to organic pollution in slow moving streams.

  alpha-mesosaprobic zone   Zone of active decomposition, that is partly aerobic and partly anaerobic, in a stream that is heavily polluted with organic wastes.

  beta-mesosaprobic zone   Zone of a stream that is moderately polluted with organic wastes.

  oligosaprobic zone   Stream reach that is slightly polluted with organic wastes and contains the mineralized products of self-purification from organic pollution with none of the organic pollution remaining.

  polysaprobic zone   Zone of a grossly polluted stream containing complex organic wastes that are decomposing primarily by anaerobic processes.

saprolite   See *saprolite* under **streambank material**.

sapropel   Neutral humus or a thick layer of old, stratified and saturated organic matter that is nearly completely mineralized.

satellite imagery   See *satellite imagery* under **remote sensing**.

saturated   Condition where all easily drained voids (i.e., interstices or pores) between soil particles are temporarily or permanently filled with water. This condition results in significant saturation if it continues for one week or more during the growing season. See also *saturated* under **water regime**.

saturated zone   (1) Area of land that is completely soaked by water where the substrate is saturated to the surface for extended periods during the growing season. (2) Zone of water with the maximum concentration of dissolved gases, elements, or other materials.

savanna river   See *savanna river* under **river**.

scan line   See *scan line* under **remote sensing**.

scanning   See *scanning* under **remote sensing**.

scarp   (1) Line of cliffs formed by the faulting or fracturing of the earth's crust or by erosion. (2) To form or cut into a steep slope. See **escarpment**.

scattered   See *scattered* under **large organic debris**.

scour   Localized erosion of substrate from the streambed by flowing water when water velocities are high.

scour chain   Steel chains implanted in the streambed to measure scour and sediment deposition within a period of time.

scour pool   See *scour pool* under *slow water*, *pool* under the main heading **channel unit**.

scour structure   See *scour structure* under **habitat enhancements**.

scrapers   See *scrapers* under **macroinvertebrate**.

scrub–shrub wetland   See *scrub–shrub wetland* under **wetlands**.

seasonal astatic   See *seasonal astatic* under **astatic**.

seasonal flow   See *seasonal flow* under *flow*.

seasonally flooded   See *seasonally flooded* under **floodplain** and **water regime**.

seasonally flooded floodplain   See *seasonally flooded floodplain* under **floodplain**.

seawater   See *seawater* under **salinity**.

secchi disk   A disk 20 or 50 cm (~ 8 or 20 in) in diameter with alternating white and black quarters that is lowered into a water column with a calibrated chain or rope used to visually measure the depth of light transparency in water.

The depth is determined directly from the calibrations on the chain or rope.

secondarily confined channel See *secondarily confined channel* under **confinement**.

secondary channel See *secondary channel* under *channel pattern* and *slow water, pool, dammed pool* under the main heading **channel unit**.

secondary current cells Generally applied to stream currents that move at right angles to the main current, primarily in meandering streams, and are responsible for the formation of concave banks and point bar deposition. Also referred to as helical flow or transverse flow.

secondary production See *secondary production* under **production**.

sediment Fragmented material from weathered rocks and organic material that is suspended in, transported by and eventually deposited by water or air.

sedimentary Substrate that is formed by the deposition of water-borne mineral fragments, organic debris, or mineral precipitates that become cemented and pressed into a solid form (i.e., rocks).

sedimentation (1) Action or process of forming and depositing sediments. (2) Deposition of suspended matter by gravity when water velocity cannot transport the bed load.

sedimentation pond See *sedimentation pond* under *pond*.

sediment budget An account of the sediment types, amounts, sources, movement, routes to specific locations, storage, and disposition of sediment in a basin.

sediment discharge Mass or volume of sediment (usually mass) passing a stream transect in a unit of time, and that is generally expressed as tons per day of suspended sediment discharge, bed load discharge, or total sediment discharge.

sediment load General term that refers to sediment moved by a stream in suspension (suspended load) or at the bottom (bed load). Sediment load is not synonymous with either discharge or concentration.

  bed load Sediment that moves on or near and frequently in contact with a streambed by rolling, sliding, and sometimes bouncing with the flow. Bedload sediments are composed of particles greater than or equal to 0.062 mm (~0.01 in) in diameter.

  bed material load Portion of the stream sediment load that is composed of particle sizes present in appreciable quantities in the streambed.

  coarse load Portion of the bed load that is more difficult to move by flowing water than sediment because it requires higher water velocities with enough power to move larger substrate materials.

  d_{50} or D_{50} Size of particle diameters that contains 50% fine sediments. The percentage of fines can be set from 1 to 100%.

  depth integration Method of sampling at all points throughout the sample depth so that a water–sediment mixture is collected proportional to the stream velocity at each point. This procedure yields a discharge weighted sample.

  fine load Portion of the total sediment load that is composed of particles smaller than the particles present in appreciable quantities in the bed material. Similar to *washload* under *sediment load*.

  suspended load Portion of the total sediment load that moves in suspension, free from contact with the streambed, and made up of small sediment particles. The density and grain size of the sediment particles are dependent upon the amount of turbulence and water velocity. Only unusually swift streams are turbulent enough or have water velocities high enough to lift particles larger than medium-sized sand from their beds. See also *bed load* and *washload* under **sediment load**.

  washload Portion of the sediment load that can be carried in large quantities and is limited only by availability in the watershed. The washload contains sediments that are finer than the smallest 10% of the bed load and usually less than 0.062 mm (~ 0.002 in) in diameter. Compare with *fine load* under *sediment load*.

sediment production zone See *sediment production zone* under *fluvial*.

sediment rating curve   A graph that illustrates the relationship between sediment discharge and stream discharge at a specific stream cross section.

sediment storage   Mineral and organic matter that is transported by a stream or river and deposited at locations where it remains in a relatively stable state.

sediment transport   Process by which individual particles of bed material are lifted from the streambed and transported by water velocity.

sediment transport rate   Mass or volume of sediment (usually mass) that passes a stream cross section in a specific unit of time.

sediment trap   See *sediment trap* under **habitat enhancements**.

sediment yield   Quantity of sediment produced from a specific area in a specified period of time.

seep   Small groundwater discharge that slowly oozes to the surface of the ground or into a stream. A seep oozes water slowly and differs from a spring that visibly flows from the ground.

seepage   (1) Movement of water through the substrate without the formation of a definite channel. (2) Loss of water by infiltration from a canal, reservoir, other water body, or field.

seepage lake   See *seepage lake* under **lake**.

seiche   See *seiche* under **wave**.

self-maintaining system   An aquatic ecosystem that can perform all of the natural ecological functions without human intervention or a dependence on engineered structures.

semipermanently flooded   See *semipermanently flooded* under **water regime**.

sensitive reservoir   Reservoir where high production rates decrease restoration or recovery. See *eutrophic* under **trophic**.

sensitive slope   Any slope that is prone to mass erosion or wasting.

sensitivity   Susceptibility of a watershed, stream, or lake to damage from natural processes or human activities.

sensor   See *sensor* under **remote sensing**.

separation bar   See *separation bar* under **bar**.

seral stages   See *seral stages* under **succession**.

serial discontinuity   Concept where dams shift biological and physical characteristics of streams and rivers from the predicted pattern related to the river continuum concept.

serpentine channel   See *serpentine channel* under **channel pattern**.

sessile   Organisms that are attached to a substrate but do not penetrate it and are unable to move about freely.

seston   All organic and inorganic material greater than 60 μm in size that is suspended in the water column.

    nannoseston   Seston that passes through a plankton net.

    net seston   Seston that does not pass through a plankton net.

settleable solids   Matter in the water column that does not stay in suspension when the water is immobile and sinks to the bottom or floats to the surface.

settling pond   See *settling pond* under **pond**.

seven day low flow (Q7L)   See *seven day low flow (Q7L)* under *flow*.

seven day/Q10   See *seven day/Q10* under *flow*.

sewage   Refuse liquids including human body wastes or wastes carried off by sewers.

shade density   Inverse of the percentage of direct light passing through crowns such that complete shading yields a value of 100%.

shallow   Term applied to water that is usually less than 2 m (less than 6.5 ft) in depth. See *shoal*.

shallow-rapid landslide   See *shallow-rapid landslide* under *landslide*.

shape index   An index of the width and depth of a stream habitat. Values less than 9 generally indicate pools, greater than 9, riffles:

$$\text{shape index} = (w/d)^{(d/d\max)}\,;$$

w = width;
d = mean depth;
d_{max} = maximum depth along a cross section.

shear stress   Force per unit area that is parallel to a surface. See also *shear stress* under **energy**.

sheen   An iridescent appearance on a water surface.

sheet   Migrating accumulations of bed load one or two grain-diameters thick that alternate between fine and coarse particles. See also *sheet* under *fast water—nonturbulent* under the main heading **channel unit**.

sheet erosion   Erosion of soil from sloping land in thin layers or sheets that may be imperceptible, particularly when caused by wind, or denoted by numerous fine rills. See also sheet erosion under **erosion**.

sheet flow   Flow of water over the ground in a more or less continuous sheet. If the flow is large, it is termed a sheet flood. See also *sheet flow* under **flow**.

sheetwash   Flow of rainwater that covers the entire ground surface with a thin film and is not concentrated in streams.

shelf   Sandbank or submerged area of rock in a water body or bedrock underlying an alluvial deposit.

Shelford law of tolerance   When one environmental factor or condition is near the limits of tolerance at either a minimum or maximum level or state, that one factor or condition will determine whether or not a species will be able to maintain itself under those specific environmental conditions.

shoal   Shallow area that is usually a sandbank, sandbar, or a rocky, swift section of stream. See *shallow*.

shoalwater substrate   Composition of the bed in a shallow (shoal) area of a river, sea, or other water body.

shooting flow   See *shooting flow* under **flow**.

shore   Land along the edge of a water body.

shoreline   Interface between land and water or the intersection of land and permanent water.

shoreline : acreage ratio   See *shoreline : acreage ratio* under **dimensions**.

shoreline development   See *shoreline development* (D_l) under **dimensions**.

shoreline length   See *shoreline length* under **dimensions**.

shredders   See *shredders* under **macroinvertebrate**.

side bar   See *side bar* under **bar**.

side channel   See *side channel* under **channel pattern**. Also, see *side channel* under *slow water, pool, dammed pool* with the main heading of **channel unit**.

signature   See *signature* under **remote sensing**.

significant wave height   See *significant wave height* under **wave**.

siliceous   Term applied to material containing silica or silica dioxide.

sill   (1) Elevated area of the bottom at the mouth of a port or harbor, or at the outlet of a water body such as a lake or estuary. (2) Bottom of a stop-log gate structure or dam crest that controls water level with flash boards. Compare with *sill* under **habitat enhancements**.

silt   (1) Fine soil that is between 0.004 and 0.062 mm (0.00002 – 0.0003 in) in diameter. (2) Also applied to a soil or substrate containing a very high proportion of silt particles. Compare with other substrate sizes under **substrate size**.

siltation   Settling of fine suspended sediments in water where water velocity is reduced.

silting   Process of depositing silt when water velocities and transport capabilities of a stream are reduced. Also applied to conditions that accompany the deposition of excessive amounts of silt.

silt load   Quantity of silt being transported in a specified quantity of water.

simple meander   See *simple meander* under **meander**.

sink   (1) Depression or low-lying, poorly drained area or hole formed where the underlying rock dissolves and waters collect in the depression or where water disappears through evaporation.

(2) Area where the input of mass or energy exceeds the output or production. (3) Location where streamflow disappears into the bed material of the stream. (4) To move downward in the water column.

sinker   Term applied to logs or large limbs that do not remain afloat in water either because of intrinsic density or through water-logging.

sinkhole   Depression, often steep-sided, that is created by subsidence where subterranean minerals or substrate dissolves and results in the collapse of an underground passage or piping.

sinking current   Downward movement of sea or lake water that has become denser through cooling or increased salinity, or moves downward as the result of an onshore wind.

sink lake   See *sink lake* under **lake**.

sink zone   See *sink zone* under **fluvial**.

sinter   General term applied to chemical sediments deposited by mineral springs.

sinuosity   (1) Ratio of channel length between two points in a channel to the straight line distance between the same two points. (2) Ratio of channel length to valley length. Channels with sinuosities of 1.5 or more are called "meandering," while those close to 1.0 are called "straight." (3) See also *sinuosity* under **dimensions**.

sinuous channel   See *sinuous channel* under **channel pattern**.

sinuous meander   See *sinuous meander* under **meander**.

site   Area described or defined by biotic, climatic, water, and soil conditions that forms the smallest planning unit with a defined boundary.

skid road   In forestry practices, any road or trail used for hauling logs from the logging site to a landing where logs are loaded on trucks or other conveyances for transport.

ski jump   See *ski jump* under **control structure**.

slab failure   See *slab failure* under **landslide**.

slack water   Quiet, still pool-like area of water in a stream usually on the side of a bend where water current is low. See *slackwater* under *slow water*,

pool, dammed pool under the main heading **channel unit**.

SLAR   See *SLAR* under **remote sensing**.

slick   (1) Glassy smooth flow of water that is sometimes used interchangeably with glide. See *glide* under *slow water* in **channel unit**. (2) A thin, shiny layer of material on the surface of the water usually referring to oil or other petroleum-based product.

slide   See *slide* under **landslide**.

sliding beads   Procedure used to measure substrate movement in a streambed that involves burying numbered beads on a cable. The beads slide to the end of the cable when substrate is dislodged from high water velocities associated with peak flows that scour the streambed. The number of beads that are dislodged provide a measure of the depth of streambed scour. See **scour**.

sliding gate   See *sliding gate* under *gate* and main heading of **control structure**.

slip erosion   See *slip erosion* under **landslide**.

slope   Incline of any part of the earth's surface. Land with an incline or oblique direction in reference to the vertical or horizontal plane. See also **gradient**.

slope break   Pattern on a slope where gradient changes abruptly.

slope failure   See *slope failure* under **landslide**.

slope processes   Mass movement by debris slides and surface wash that results in transport of fine sediments downslope by overland flow.

slope stability   Measure of slope resistance to erosion, slumping, sliding, or other unstable conditions.

slope wash   Motion of water and sediments down a slope caused by sheet flow.

sloping gully side   See *sloping gully side* under **gully side form**.

slough   (1) Low swamp or swamp-like area in a marshy or reedy pool, pond, inlet, or backwater with marsh characteristics such as abundant vegetation. (2) Channel where water flows sluggishly or slowly through low swampy ground on a delta or floodplain. (3) Marshy tract located in a shallow, undrained depression or a sluggish creek in a bottomland. (4) Tidal channel in a salt marsh. (5) Lower reach of a tributary that has been ponded by sediment and debris at the confluence with the main channel.

slow water   See *slow water* under **channel unit**.

sludge   Deposit of a semifluid mass such as mud, ooze, sediment, or organic matter in the bottom of a water body.

sluggish flow   See *sluggish flow* under **flow**.

slump   See *slump* under **landslide**.

slump-earthflow   See *slump-earthflow* under **landslide**.

small bole   See *small bole* under **large organic debris**.

small impoundment   Small reservoir (generally 8 ha [20 acres] or less in surface area) used for water storage and control on a stream. See **reservoir**.

small mountain lake   See *small mountain lake* under **lake**.

small-sporadic deep-seated failures   See *small-sporadic deep-seated failures* under **landslide**.

snag   (1) Standing dead tree. (2) Submerged fallen tree in a stream, sometimes with an exposed or only slightly submerged tree top. (3) See also *snag* under **large organic debris**.

snagging   Removing or cutting snags on land or in water.

soda lake   See *soda lake* under **lake**.

softwater   Freshwater with low alkalinity, conductivity, or salinity that is generally found in areas with sandstone substrate or headwater streams. Compare with **hardwater**.

soil   Portion of the earth's surface that consists of earth, disintegrated rock, and humus and that is capable of supporting vegetation. See **earth, ground**.

soil creep   Gradual downslope movement of soil by gravity.

soil drainage   Pattern of water drainage from soils, generally applied to saturated soil.

soil erosion   Removal of soil through erosion by wind, precipitation, surface water, or other natural processes.

soil pore   Area, interstice, or space within soil that is occupied by either air or water where the degree of porosity is dependent upon the arrangement of individual soil particles.

soil water potential   Amount of work required to transport a given quantity of water from the surface into groundwater.

solar arc   See *solar arc* under **solar radiation**.

solar radiation   Electromagnetic energy from the sun at all wavelengths. More particularly, radiation with wavelengths between 0.2 and 4.5 μm that emanate from the sun and can be measured by instruments (e.g., actinometer, pyranometer, radiometer, or solimeter). The fraction of incident light or electromagnetic radiation that is reflected by a surface or body is known as albedo. Net radiation is the algebraic sum of the upward and downward vertical components of long- and short-wave radiation.

arc of the sun   Change in the angle of the sun on a given day in degrees from when sunlight first strikes water. The arc of the sun on August 1st at the same location is used as a standard.

direct solar radiation   Radiation that reaches a water surface in an unobstructed straight line.

incident light   Visible light reaching a water surface.

reflected solar radiation   Radiation that does not penetrate a water surface but is reflected from the surface.

refracted solar radiation   Radiation that penetrates a water surface and is either bent or deflected from its original path.

solar arc   Measure of canopy angle in degrees that is based on the measurement of the angles formed from the line of sight to the visible horizon.

total solar radiation   Sum of direct, reflected, and refracted radiation reaching a given point.

solifluxion, solifuction Slow, downhill flow of soil or soil layers saturated with water that is typical of sites subjected to periods of alternate freezing and thawing. Solifluxion over frozen ground is termed gelifluxion that may develop into a sudden mass flow of mud and earth.

soligenous fen See *soligenous fen* under ***wetlands***.

solitron A single, isolated peak or trough of a wave in water.

solum Upper part of a soil profile that is influenced by plant roots.

solute Substance that will dissolve or go into solution.

solution lake See *solution lake* under ***lake***.

sonar See *sonar* under ***remote sensing***.

sorted Geological term pertaining to the variability of particle sizes in a clastic sediment or sedimentation rock. Materials with a wide range of particle sizes are termed poorly sorted while materials with a small range of sizes are termed well sorted.

sorting coefficient Measure of the distribution or variability of particle sizes in substrate that is usually expressed as the square root of d_{75}/d_{25}. The terms d_{75} and d_{25} are diameters where 75% and 25% of the cumulative size-frequency distributions are larger than a given size. A substrate with a large sorting coefficient is termed well sorted.

sound Water body that is usually broad, elongated, and parallel to the shore between the mainland of a continent and one or more islands.

sounding Process of measuring the depth of water, as in a lake or reservoir, either with a sounding line (chain or cord with gradations) or by transmitting a sound wave into water to reflect off the bottom and read on a depth finder.

spawning box See *spawning box* under ***habitat enhancements***.

spawning marsh See *spawning marsh* under ***habitat enhancements***.

spawning platform See *spawning platform* under ***habitat enhancements***.

spawning reef See *spawning reef* under ***habitat enhancements***.

spawning substrates Substrates of suitable size and composition in rivers, lakes, or other water bodies that are used by fish and and other aquatic animals for deposition of eggs and sperm.

specific gravity Ratio that denotes the density of an object or fluid compared to the same volume of distilled water at 4°C.

specific reach See *specific reach* under ***reach***.

spectral signature See *spectral signature/spectral relectance curve* under ***remote sensing***.

spiling See *spiling* under ***habitat enhancements***.

spilling breaker See *spilling breaker* under ***wave***.

spillway See *spillway* under ***control structure***.

spit Narrow strip of land that projects into a water body.

splash apron Concrete or rock structure placed at the outlet of a culvert, drop structure, channel, or ford to intercept water and reduce velocity to prevent scouring.

splash dam Temporary or permanent structure in a stream channel used to store logs and water until sufficient water is present from precipitation, runoff, and storage to transport the logs downstream when the splash dam is opened.

splash erosion See *splash erosion* under ***erosion***.

SPOT See *SPOT* under ***remote sensing***.

spring Site where groundwater flows naturally from a rock or soil substrate to the surface to form a stream, pond, marsh, or other type of water body.

    cold spring Spring with a mean annual water temperature that is appreciably below the mean annual atmospheric temperature for a specific area.

    hot spring Thermal spring with a mean water temperature that exceed the normal temperature of a human body (37°C or 98.6°F).

spring breakup (1) Breakup of ice on rivers and lakes during the spring thaw. (2) Period in spring when snow and ice that are formed during winter are melting. (3) In some parts of the United States, this term refers to the "mud season" when earth or clay roads may be impassable.

springbrook   Short, spring-fed stream with substrates of organic mud and sand that often contains thick growths of watercress.

spring creek   Stream that derives most of its flow from a spring and is characterized by a relatively constant flow and water temperature.

spring overturn   See *spring overturn* under *stratification*.

spring source area   Lands that contribute water to a spring by infiltration and percolation through the soil or other substrates.

sprungschicht   See *metalimnion* and *thermocline* under *stratification*.

stability   Ability of a bank, streambed, or slope to retain its shape and dimensions when exposed to high streamflows and varying temperature conditions.

stability rating   An index of the resistance or susceptibility of the stream channel and banks to erosion.

stable bank   See *stable bank* under *bank stability*.

stable debris   See *stable debris* under *large organic debris*.

stable flow   See *stable flow* under *flow*.

stage   (1) Elevation of a water surface above or below an established reference point. (2) Quantification of a discharge expressed as a percent of mean annual discharge or some other reference flow. (3) Depth of water at any point in a stream that is generally calibrated so that discharge can be estimated with a weir.

stage class   See *stage class* under *succession*.

stagnant   Layer of inert water with little or no circulation of water and low dissolved oxygen.

stagnation point   Point at the leading edge of an object where the water velocity of the oncoming flow is zero where the water collides with the object.

stagnation pressure   Pressure differential between a zone of high velocity water flow and a zone of lower velocity such as along rock riprap. Such pressure differential or stagnation pressure is often sufficient to dislodge large, heavy objects such as boulders and large rocks.

stake bed   See *stake bed* under **habitat enhancements**.

standard fall velocity   Average velocity that a particle would finally attain if falling in quiescent distilled water at a temperature of 24°C.

standard sedimentation diameter   Diameter of a sphere that has the same specific gravity and the same standard fall velocity as a given particle.

standing crop   Quantity of living organisms present in the environment at a given time that is usually expressed as the dry total weight of biomass of a specific taxon or community. Generally synonymous with **standing stock** but often refers to the harvestable portion of the **standing stock**. See **standing stock**.

standing stock   Dry total weight of biomass of a specific taxon or community of organisms that exists in an area at a given time. See **biomass**.

standing timber   Trees, usually dead, that were left uncut in a reservoir basin before impoundment to serve as habitat for fish and invertebrates.

standing water   Water that remains in one location such as a marsh, pond, lake, or swamp. See **lentic**.

standing wave   See *standing wave* under **wave**.

static head   Distance from a standard datum (artificially defined reference point) of the water surface on a column of water that can be supported by the static pressure at a given time.

station   (1) Exact place of occurrence of an individual or species within a given habitat. (2) Permanent or semi-permanent sample area. (3) A circumscribed area that contains all the environmental conditions required by an individual or group of individuals.

status   See *status* under **trophic**.

steady flow   See *steady flow* under **flow**.

steep bank   See *steep bank* under **streambank**.

stem flow   See *stem flow* under *flow*.

stenobathic   Refers to an organism that is restricted to living at a certain depth in a water column.

stenohaline   See *stenohaline* under *salinity*.

stenosaline   See *stenosaline* under *salinity*.

stenotherm   Organisms that have a narrow temperature tolerance.

stenotypic organism   An organism with a narrow range of tolerance to a particular environmental factor.

step run   See *step run* under *fast water—turbulent* under the main heading *channel unit*.

still   Water is considered to be still when it is motionless, free from turbulence, without waves or perceptible current.

stilling basin   Deep pool located in or below the spillway of a dam that dissipates the energy of the water in the spillway.

stilling pond   See *stilling pond* under *pond*.

stocking   Release of bird, fish, or wildlife species into a given habitat that were obtained through captive propagation or were captured from the wild elsewhere.

stone   Naturally formed hard substance consisting of mineral or earth materials such as rock formations, weathered rock in the form of boulders, or gravel (pieces of rock that have become rounded from scouring in a streambed). Compare with other substrate sizes under *substrate size*.

stop log   See *stop log* under *control structure*.

storage   (1) Water that is artificially impounded in surface or underground reservoirs for future use. (2) Water that is naturally detained in a drainage basin as groundwater, channel storage, and depression storage. The term "drainage basin storage" or simply "basin storage" is sometimes used in reference to the total amount of naturally stored water in a drainage basin.

storage coefficient   Coefficient that expresses the relation of storage capacity in a reservoir to the mean annual flow of a single stream or all streams that are direct tributaries to the reservoir.

storage ratio   Net available water storage divided by the mean annual flow for a basin within a period of one year.

storage reservoir   Reservoir designed to retain water during peak water periods for release and use downstream at another time when streamflows are low. See *reservoir, storage*.

storm event   Major episode of atmospheric disturbance that is often associated with heavy precipitation, lighting, and thunder.

storm flow   See *storm flow* under *flow*.

story   One of several distinct layers of plant growth such as tall trees, large shrubs, low shrubs, and ground cover.

straight   See *straight* under *slow water, scour pool* under *channel unit*.

straight channel   See *straight channel* under *channel pattern*.

straight meander   See *straight meander* under *meander*.

strait   Narrow passage of water connecting two larger water bodies.

strath terrace   Terrace composed of ancient alluvial material that is deposited as a mantle over a base of bedrock.

stratification   Arrangement of water masses into distinct, horizontal layers that are separated by differences in density associated with water temperature and dissolved or suspended matter.

bathylimnion   Deepest part of a lake that is located below the clinolimnion.

chemocline   Density gradient, or pycnocline, from differences in salt concentration.

clinograde   Oxygen profile in which the hypolimnion has less dissolved oxygen than the epilimnion.

clinolimnion   Layer or region of a water body where the rate of heating decreases exponentially.

destratification   Process that interrupts the boundary between water strata and induces mixing between strata.

stratification layers (adapted from Thorp and Covich 1991)

discontinuity layer   See *thermocline* under *stratification*.

epilimnion   Uppermost layer of water in a lake characterized by an essentially uniform temperature where relatively thorough mixing occurs from wind and wave action to produce a less dense but oxygen-rich layer of water. In a thermally stratified lake the epilimnion extends from the water surface down to the metalimnion.

fall overturn   Physical phenomenon that involves the thorough mixing of water that occurs in temperate-zone water bodies during the fall season. The sequence of events leading to the fall overturn includes: (a) cooling and increased density of surface waters producing convection currents from top to bottom; (b) circulation of the total water volume by wind actions and density differences resulting in a uniform water temperature that allows complete mixing of nutrients and chemicals throughout the entire water mass.

hypolimnion   Poorly oxygenated and illuminated lower layer or region in a stratified lake that extends from the metalimnion to the bottom and is essentially removed from major surface influences. Water in the hypolimnion is denser and colder than strata higher in the water column.

inverse stratification   Water body with colder water in a stratum over warmer water.

mesocline   See *thermocline* under *stratification*.

mesolimnion   Term used in place of thermocline. See *thermocline* under *stratification*.

metalimnion   Stratum between the epilimnion and hypolimnion that exhibits a marked thermal discontinuity with a temperature gradient equal to or exceeding 1°C per meter. See *mesocline, thermocline*.

mixolimnion   Upper strata of a lake that exhibits periodic circulation. See *epilimnion* under *stratification*.

monimolimnion   Deeper stratum or layer of a lake that remains perennially stagnant and rarely circulates, especially the layer below the chemocline in a meromictic lake.

negative heterograde   Term applied to a vertical profile of water with minimum oxygen levels.

orthograde   Oxygen profile where the oxygen level below the epilimnion remains at or near saturation.

overturn   Period of mixing or circulation of water in a previously thermally stratified water body. See *fall overturn* under *stratification*.

positive heterograde   Profile where an oxygen profile remains well above saturation.

pycnocline   Layer or region in saltwater where a marked change occurs in the density of a water column that acts as a partial barrier to exchange between the upper and lower water columns.

spring overturn   Physical phenomenon that may involve the thorough mixing of water in temperate-zone water bodies during the early spring. The sequence of events leading to spring overturn includes: (a) melting ice cover; (b) warming surface waters; (c) changing densities in surface waters that produce convection currents from top to bottom; and

(d) circulation of the entire water volume by wind action, resulting in a uniform water temperature that allows complete mixing of nutrients and chemicals throughout the entire water mass.

sprungschicht   Term that is synonymous with thermocline or metalimnion.

thermal stratification   Vertical temperature stratification in north temperate lakes resulting in: (a) virtually uniform water temperature in the epilimnion; (b) rapid and marked gradient change in temperature with depth in the metalimnion; and (c) cold and nearly uniform water temperature in the hypolimnion, from the bottom of the metalimnion to the bottom of a water body.

thermocline   Stratum between the epilimnion and hypolimnion that exhibits a marked temperature gradient equal to or exceeding 1°C per meter. Synonomous with mesolimnion or metalimnion.

turnover   Refers to the thorough mixing of water in a lake by wind action that occurs when density differences of a thermally stratified water column disappear and temperatures become uniform. See fall turnover and spring turnover.

stratified   Refers to a series of water layers that form from density differences of water temperatures. See *thermocline* under **stratification**.

stratified flow   Layered flow that results from a difference in density due to temperature, dissolved, or suspended materials between the inflowing and receiving water.

stratified lake   See *stratified lake* under **lake**.

stratified stream segment   Portion of a stream that is relatively homogeneous based on geomorphology, streamflow, geology, and sinuosity. This term also refers to a series of short reaches with a common morphology.

streaks   Surface areas parallel to direction of the wind that coincide with lines of surface convergence and downward movement of water. See also *langmuir circulation* under **wave**.

stream   Natural water course containing flowing water, at least part of the year, together with dissolved and suspended materials, that normally supports communities of plants and animals within the channel and the riparian vegetation zone.

alluvial stream   Stream where the form of the streambed is composed of appreciable quantities of sediments that are transported and deposited in concert with changes in streamflow.

beaded stream   Stream connecting a series of small ponds or lakes.

beheaded stream   Stream that has been separated from a portion of its headwater tributaries. See *stream piracy*.

centripetal stream   Streams that converge in the central part of a basin.

consequent stream   Stream that flows in the same direction as local geologic strata.

continuous stream   Stream where the flow along its course is not interrupted in space or time.

distributary   Division of stream channels, as on a delta or alluvial fan, that flow away from the main channel, usually into a larger stream, lake, or other receiving water body. Distributary streams form where deposition exceeds erosion.

entrenched stream   Stream that has eroded into the substrate and is confined by walls resistant to erosion.

gaining stream   Stream or stream reach that receives water from the zone of saturation.

graded stream   Stream that has achieved a state of equilibrium between the rate of sediment supply, transport, and deposition throughout long reaches.

headwater stream   Stream that has few or no tributaries, and has steep, incised channels that are often associated with active erosion, seeps, and springs. Headwater streams are referred to as first order streams. See *seep, spring, stream order*.

incised stream   Stream that has, through degradation, cut its channel into the bed of a valley.

insulated stream   Stream or stream reach that neither contributes to nor receives water from the zone of saturation because it is separated by an impermeable bed.

interrupted stream   Stream without a continuous flow where reaches with water may be perennial, intermittent, or ephemeral.

losing stream   Stream or stream reach that contributes water to the zone of saturation.

lowland stream   Stream that flows across low gradient terrain, has a bed composed of fine substrate materials, and is located in the lower part of a drainage network downstream from mountains.

mature stream   See *graded stream* under **stream**.

meltwater stream   Channelized flow of glacial melt water.

middle stream   Refers to a stream reach that is located in the center of a stream course between headwaters and the mouth, has beds of diverse substrates, variable habitat patterns, and usually has equilibrium or balance between scouring and deposition. See *deposition, scour*.

obsequent stream   Stream that flows in a direction opposite of the general trend in local geologic strata.

parallel stream   Stream that flows in close proximity to and in the same direction as another stream but separated by a divide.

perched stream   Stream that may be classified as either "losing" or "isolated" and is separated from the underlying groundwater by a zone of aeration. See *losing stream* and *isolated stream* under **stream**.

pinnate stream   Stream pattern characterized by a series of small tributaries distributed along the stream gradient of the main stem.

pooled channel   An intermittent stream with significant surface pool area and without flowing surface water that is supplied by groundwater.

radial   Pattern of stream channels flowing out from a central point such as a volcanic cone.

rectangular stream   A system of streams in which each straight segment of stream takes one of two characteristic perpendicular directions that follow perpendicular landforms.

superimposed stream   A stream whose course, once established, is maintained by erosion cutting deeper into the landform.

trellis stream   Stream pattern in which tributary streams join the main stream at or near right angles and are fed by elongated secondary tributaries parallel to the main stream so that the whole system resembles a vine on a trellis.

streambank   Ground bordering a channel above the streambed and below the level of rooted vegetation that often has a gradient steeper than 45° and exhibits a distinct break in slope from the stream bottom. The portion of the channel cross section that restricts lateral movement of water during normal streamflow. Right and left banks are determined while looking downstream.

concave bank   Bank that is indented such that the top and bottom of the bank are higher and is often characterized by bank erosion. Generally, a concave bank is located on the outside of a river curve or bend of a meandering stream.

convex bank   Bank that is inverted such that the top and bottom of the bank are lower and is often characterized by sediment deposition at a point bar. Generally, a convex bank is located on the inside bank on a river curve or bend, especially on a meandering stream.

cut bank   Streambank that is actively eroding and has a steep face.

flat bank   Streambank where the riverbed slopes gently to the level of rooted vegetation.

lower bank   Bank that is periodically submerged between the normal high water line to the water's edge during the summer low flow period.

repose bank   Bank with an angle of repose (usually 34–37°) in unconsolidated material.

steep bank   Bank that is nearly vertical and is consolidated by the cement action of minerals, compaction, and roots from riparian vegetation.

undercut bank   Bank with a cavity below the water line that is maintained by scour from substrates and high water velocities. See *scour*.

unstable bank   Streambank boundary of the channel that is actively failing through erosion or slumping, that is recognized by clumps of sod and earth along the base of the bank, and that is expressed as a percentage of the total length of both banks for the reach. See *erosion* and *earth slump* under **landslide**.

upper bank Portion of a bank in the topographic cross section of the channel from the break in the general slope of the surrounding land to the normal high water line.

streambank material Substrates that compose banks along stream courses. The following terms classify bank material according to composition, origin, and method of formation.

anthropogenic Materials created or modified by humans, including those associated with mining of minerals, waste disposal, and erosion control. See *revetment, riprap*.

bedrock Rock outcrop or rock covered by a thin mantle (less than 10 cm) of consolidated material.

colluvial Product of mass movement of materials (usually angular and poorly sorted) that reached their present position by direct influence of gravity such as slides or talus slopes. See *landslide*.

eolian Materials (usually silt or fine sand) that are transported and deposited by wind.

fluvial Materials (usually rounded, sorted into horizontal layers, and poorly compacted) that are transported and deposited by streams and rivers.

ice Frozen water from seeps or glaciers, or atmospheric conditions. See *glacier, seep*.

lacustrine Fine-textured sediments that have settled to the bottom from suspension in bodies of standing freshwater or that have accumulated at the margins of lakes or reservoirs through wave action. May be fine textured with repetitive layers.

marine Sediments that have settled from suspension in brackish or saltwater of estuaries or oceans.

morainal Poorly sorted (angular to subangular) material transported in front of, beneath, beside, or within a glacier and deposited directly by the glacier. May be highly compacted and have significant clay content.

organic Materials resulting from vegetative growth, decay, and accumulation in closed basins or on gentle slopes where the rate of accumulation exceeds that of decay.

saprolite Weathered bedrock that decomposed in situ, principally by chemical and weathering processes.

undifferentiated Multiple layers of different types of material.

volcanic Unconsolidated volcanic or igneous (pyroclastic) sediments that accumulate from volcanic eruptions or fine volcanic dust carried by winds and deposited some distance from the volcano.

streambank stability Index of firmness or resistance to disintegration of a bank based on the percentage of the bank showing active erosion and the presence of protective vegetation, woody material, or rock. See *stable* and *unstable* under *bank stability*, and *cut bank* and *unstable bank* under *streambank*.

streambed Substrate plane, bounded by banks, of a stream bottom. Also referred to as the stream bottom.

stream capacity (1) Total volume of water that a stream can transport within the high water channel. (2) Maximum sediment load a stream can transport at a given velocity and discharge.

stream capture Upstream connection of one stream by erosion into the drainage basin of another stream that results in changes of drainage patterns. See *stream piracy*.

stream channel (1) Long, narrow depression shaped by the concentrated flow of a stream and covered continuously or periodically by water. (2) Bed and banks formed by fluvial processes where a natural stream of water runs continually or intermittently. See *channel*.

stream classification Systems used to group or identify streams possessing similar features using geomorphic structure (e.g., gradient and confinement), water source (e.g., spring creek), associated biota (e.g., trout zone), or other characteristics. A hierarchical classification. Two approaches are commonly used: a management-related classification that is based almost entirely on value to fish populations, and a geomorphic-habitat classification system.

stream corridor Perennial, intermittent, or ephemeral stream and riparian vegetative fringe that occupies the continuous low profile of the stream valley.

stream density Abundance of streams that is expressed as kilometers of stream per square kilometer of landscape or terrain. Synonymous with *drainage density*. See *landscape, terrain*.

stream discharge See *stream discharge* under *discharge*.

stream–estuary ecotone Transitional area from a stream mouth and lower limit of marsh vegetation that extends to the upper limit of tidal influence. See *ecotone*.

streamflow See *streamflow* under *flow*. See also *discharge*.

stream–forest ecotone Area of a stream that is directly influenced by riparian vegetation, including the streambank and upland area adjacent to the stream. Its size depends on stream width, type of vegetation, and physical characteristics of the adjoining uplands. See *ecotone, streamside management zone*.

stream frequency The number of streams per square kilometer of area.

streamline Direction of water movement at a given instant.

streamline flow See *streamline flow* under *flow*.

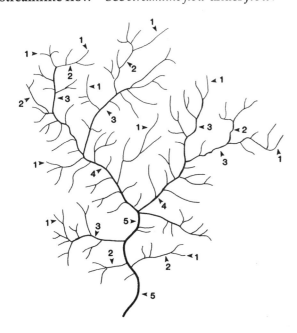

stream order (from Helm 1985)

stream order Hierarchical ordering of streams based on the degree of branching. A first-order stream is an unforked or unbranched stream.

Two first-order streams flow together to form a second-order stream, two second orders combine to make a third-order stream, etc.

stream pattern See *channel type*.

stream piracy Transfer of a stream from one basin to another as a result of geomorphic changes, usually by erosion through a common divide. See *stream capture*.

stream power Energy or ability of a stream to move substrates and scour streambanks that is based on gravity, slope, discharge, and water velocity. See also *stream power* under *energy*.

stream profile Graphical presentation of elevation versus distance of a stream channel or shape of a cross section. In open channel hydraulics, it is a plot of water surface elevation against channel distance.

stream reach See *reach*.

stream shore water depth Water depth at a stream shoreline or at the edge of a bank overhanging a shoreline.

streamside management zone The land, together with the riparian vegetation, that is in immediate contact with a stream and sufficiently close to have a major influence on the total ecological character and function of the stream. See *stream–forest ecotone*.

stream surface shading Percentage of the stream surface area that is shaded.

 foliar shading Shading attributable to riparian vegetation.

 nonfoliar shading Shading from debris, undercut banks, and surrounding terrain.

stream transport capability Ability of a stream to move sedimentary materials and organic material during periods of peak flow. This ability is influenced by water volume (i.e., discharge), water velocities, channel slope, timing of the peak flows, and the presence or absence of obstructions in the channel.

stream width See *wetted width* under *dimensions*.

structure (1) Any object, usually large, in a channel that influences streamflow. (2) Features that create a diversity of physical habitat within a stream, lake, or reservoir. (3) Organization of taxa into various functional or trophic groupings in a biological community. Compare with *structure* under **habitat enhancements**.

stump field   Area in the bottom of an impoundment where stumps remain after trees were cut and removed before impoundment.

subbasin   Surface area of a watershed drained by a tributary to a larger stream that is bounded by ridges or other hydrologic divides and is located within the larger watershed drained by the larger stream. Also referred to as subdrainage.

subdrainage   See *subbasin*.

sublittoral   See *sublittoral* under **littoral**.

submerged   Refers to under a water surface.

submerged macrophyte   See *submerged macrophyte* under **macrophyte**.

subsidence   Lowering of surface elevations caused by loss of support and subsequent settling or caving of substrate strata.

substrate   (1) Mineral and organic material forming the bottom of a waterway or water body. (2) The base or substance upon which an organism is growing. Also referred to as substratum.

substrate size   The following table includes the average diameter of various substrates in millimeters and inches.

| Name of particle | Diameter of particle | |
	Millimeters	Inches
Large boulders	>1,024	40–160
Small boulders	256–1,024	10–40
Stone	256–600	10–24
Rubble (large cobble)	128–256	5–10
Cobble (small cobble)	64–128	2.5–5
Pebble	2–64	0.08–2.5
Coarse gravel	32–64	1.3–2.5
Fine gravel	2–32	0.08–1.3
Sand	0.062–2.0	–
Silt	0.004–0.062	–
Clay	<0.004	–

subsurface flow   Water that moves horizontally below the earth's surface. See also *subsurface flow* under *flow*.

subsurface inflow   Water moving horizontally through the upper soil layers into a water body.

subsurface outflow   Water moving horizontally through the upper soil layers away from a water body.

subsurface runoff   Term for the portion of runoff that percolates through the soil by gravity as groundwater before emerging to the surface as seepage or springs.

subterranean stream   Part of a stream reach that flows underground.

succession   Changes in species composition of plants and animals in an ecosystem with time, often in a predictable order. More specifically, the gradual and natural progression of physical and biological changes, especially in trophic structure of an ecosystem, toward a climax condition or stage.

   climax succession stage   Culminating stage in plant succession for a given site where the vegetation has reached a highly stable condition.

   early succession stage   Stage in forest or other plant community that includes species that colonize early, or a stage of young growth including seedlings, saplings, and pole-sized trees.

   late succession stage   Plant community that has developed mature characteristics. Also referred to as the climax stage.

   mid-successional stage   Succession in a plant community between the early colonizers and transition to a mature community.

   seral stages   Series of relatively transitory plant communities that develop with ecological succession from bare ground to the climax stage.

   stage class   Any distinguishable phase of growth or development of a population or community.

summer heat income   See *summer heat income* under **heat budget**.

summerkill   Complete or partial dieoff of a fish population during summer when extended cloud cover prevents sunlight from allowing photosynthesis of plants. The death of rooted aquatic plants depletes dissolved oxygen in warm water and fish die by suffocation or from toxins produced by certain species of decaying algae.

summer pool   Water level in a reservoir, usually maintained at a stable level during the summer months. Compare with *recreational pool*.

sump   A pit, well, or depression where water or other liquid is collected.

superimposed stream   See *superimposed stream* under *stream*.

supervised classification   See *supervised classification* under *remote sensing*.

supervised training   See *supervised training* under *remote sensing*.

supralittoral   See *supralittoral* under *littoral*.

surf   Waves or swells that break upon the shore of a water body. Generally this term applies to ocean environments.

surface   Interface between water and the atmosphere.

surface area   Area of a water body at the interface between water and the atmosphere.

surface creep erosion   See *surface creep erosion* under *erosion*.

surface elevation   Height of the surface above mean sea level.

surface erosion   See *surface erosion* under *erosion*.

surface film   Surface tension of water due to air–water (molecular) interactions that provides support for certain plants and animals so they can live at the air–water interface.

surface flow   See *surface flow* under *flow*.

surface impoundment   Natural topographic depression, artificial excavation, or dike arrangement containing water that is constructed above, below, or partially in the ground (or in navigable waters) and may or may not have a permeable bottom and sides.

surface runoff   Portion of the runoff that flows over the surface without infiltrating into the groundwater.

surface seiche   See *surface seiche* under *wave*.

surface water   Standing water above the substrate or water that flows exclusively across a land surface and includes all perennial and ephemeral water bodies.

surface water inflow   Water flowing into a water body from one or more stream channels.

surface water outflow   Water flowing out of a water body in one or more stream channels.

surface wave   See *surface wave* under *wave*.

surfactant   Chemical agent that improves the emulsifying, dispersing, spreading, or wetting abilities of other chemicals such as herbicides and pesticides.

surge   An episode of uneven flow and strong momentum such as a swell due to a sudden change in pressure or a rapid destabilizing physical event. See *swell*.

survival flow   See *survival flow* under *flow*.

suspended load   See *suspended load* under *sediment load*. Compare with *bed load* and *washload* under *sediment load*.

suspended sediment discharge   Quantity, usually expressed as mass or volume, of suspended sediment passing a stream cross section in a given unit of time.

suspended sediments   Sediments that are carried in suspension in the water column by turbulence and water velocity or by Brownian movement so that they are transported for a long time without settling to the bottom. See *suspended load*.

suspended solids   Particles of unfiltered, undissolved solid matter such as wood fibers or soil that are present in water. See *total dissolved solids (TDS)*.

suspension erosion   See *suspension erosion* under *erosion*.

suspensoids   Colloidal particles that remain in suspension under most conditions and combine or react with liquid only to a limited extent.

sustained (cruising) speed   See *sustained (cruising) speed* under *swimming speed*.

swale   A moist or marshy depression or topographic low area, particularly in prairies.

swamp   See *swamp* under *wetlands*.

swash   Landmark rush of water from a breaking wave up the slope of a beach.

swath width   See *swath width* under *remote sensing*.

sweeper log   See *sweeper log* under *large organic debris*.

swell   See *swell* under *wave*.

swimming speed   Speed that fish or other aquatic organisms travel when swimming that varies from essentially zero to over six m/s (19.7 ft/s), depending upon species, size, and activity. Swimming speed of fish is often expressed as body length per second. Three categories of swimming performance are generally recognized:

> **burst (darting) speed**   Speed that a fish can maintain for a very short time, generally 5–10 seconds, without fatigue. Burst speeds are used by fish in feeding, escape from predators, or to pass barriers such as cascades or falls during migration, and represent the maximum swimming speed for a species.

> **prolonged speed**   Speed that a fish can maintain for a prolonged period of time (minutes but usually less than 1 hr) that ultimately results in fatigue. A fish is under some degree of stress when it swims at its prolonged speed. Prolonged speeds in fish typically involve anaerobic metabolism.

> **sustained (cruising) speed**   Speed that a fish can maintain for an extended period of time (hours) without fatigue or stress during normal movements between two sites. Cruising speeds in fish typically involve aerobic metabolism.

swimming velocity   See *swimming speed*.

swirling flow   See *swirling flow* under *flow*.

synergism   Interaction of two or more substances (e.g., chemicals) such that the action of any one of them on living cells or tissues is increased. Compare with *antagonism*.

synthetic aperture radar (SAR)   See *synthetic aperture radar (SAR)* under *remote sensing*.

synusia   Any component of a community that is composed of one or more species, belonging to the same life-form, having similar environmental requirements, and occurring in similar habitats.

system   Regularly interacting or interdependent group of items or things forming a unified whole. In stream applications, it includes a watershed or basin. For examples, see *basin, drainage, ecosystem, watershed*.

▶ t

taiga   Northern, subarctic coniferous forest (composed of sparse stands of small spruces and firs) in Asia, Europe, and North America that is typically open or interspersed with bogs and forms a transition zone between denser forest to the south and tundra to the north.

tail   Transition between habitat types that is usually shallow where the water velocity increases (e.g., the downstream section of a pool, glide, or other habitat type). Synonymous with tailout or flat.

tailings   Mining waste from screening or processing mineral ore. Tailings may contain suspended solids, heavy metals, radioactive materials, acids, and other contaminants that can leach into a water course.

tailrace   (1) Channel with highly turbulent water, usually confined by concrete or riprap, in the tailwater of a reservoir. (2) Channel that carries water from a water wheel.

tailwater   Flowing water below a dam that is released from an upstream impoundment. Often releases from the hypolimnion in the reservoir provides clear, cold water in the tailwater that can support coldwater sport fisheries.

talus   (1) Slope with numerous, loosely aggregated rocks. (2) The sloping accumulation of rock fragments at the base of a cliff.

tank   See *tank* under *pond*.

taxon   Any formal taxonomic unit or category of organisms (e.g., species, genus, family, order). The plural of taxon is taxa.

tectonic lake   See *tectonic lake* under *lake*.

telluric water   Surface water derived from sources other than direct precipitation on a site (e.g. seep or spring). See *seep, spring*.

temporarily flooded See *temporarily flooded* under *water regime*.

tenaja Pools in seasonal streams that may support a flora similar to vernal pools upon desiccation.

terminal moraine See *terminal moraine* under *moraine*.

terrace Relatively level or gently inclined land surface that is elevated above an active stream channel in a steplike arrangement of a slope or lake bed. Terraces are remnants of floodplains or perched shorelines at low water levels. See *bench*.

terrace tributary See *terrace tributary* under *tributary*.

terracing Dikes constructed along the contour of agricultural land to contain runoff and sediment, thereby reducing erosion. See *dike*.

terrain Comprehensive term describing the landscape with respect to its features. See *landscape*.

terrestrial Belonging to, or living on, the ground or earth.

terrigenous sediments Sediments produced from soil, ground, earth, or weathered rock that are derived directly from the neighboring land.

tetrapod See *tetrapod* under *habitat enhancements*.

texture Size, shape, and arrangement of particles in a substrate.

thalweg Path of a stream that follows the deepest part of the channel.

thalweg depth Vertical distance from the water surface to the deepest point of a channel cross section.

thalweg velocity See *thalweg velocity* under *velocity*.

thaw lake See *thaw lake* under *lake*.

thematic data See *thematic data* under *remote sensing*.

thematic mapper See *thematic mapper* under *remote sensing*.

theme See *theme* under *remote sensing*.

thermal Related to or caused by heating or warm temperatures.

thermal bar Narrow transition zone of nearly 4°C in the vertical isotherm between an open water mass and the stratified area of a water body.

thermal refuge Zone in a water body that maintains oxygenated water and temperatures adequate for fish survival.

thermal stratification See *thermal stratification* under *stratification*.

thermocline See *thermocline* under *stratification*.

thermokarst Depression created in permafrost by melting of ground ice with a subsequent settling of the soil.

thermokarst lake See *thermokarst lake* under *lake*.

throughfall All the precipitation that eventually reaches a forest floor including direct precipitation and drip from foliage but minus stem flow.

tidal flat Level land that is regularly inundated by ocean tides, often with muddy substrate.

tidal inlet An opening along the shoreline where water extends at high tide.

tidal marsh See *tidal marsh* under *wetlands*.

tide Alternate rising and falling of the surface of an ocean (including bays and gulfs connected to an ocean) that generally occurs twice daily from the gravitational pull of the moon and sun on the earth.

till Unmodified substrate material deposited by glaciers and ice sheets.

timber crib See *timber crib* under *habitat enhancements*.

tin whistle See *tin whistle* under *control structure*.

tire reef See *tire reef* under *habitat enhancements*.

toe The base of a slope along a bank or other geographic feature where a gentle incline changes abruptly to a steeper gradient.

toe of the bank The point at the base of a streambank where the bank becomes more level as it forms the channel bed.

toe undercutting   Erosion by a stream at the toe of an underwater slope that may result in bank failure.

toe width   The width of the exposed toe along a slope.

tolerance   Relative capability of an organism to endure or adapt to unfavorable environmental conditions.

tolerance association   Association of organisms capable of withstanding adverse environmental conditions within a habitat. This association is often characterized by a reduction in the number of species and, in the case of organic pollution, an increase in individuals representing certain species.

tolerance limit   Concentration of a substance that can be endured by an organism for a specified period of time.

tolerance quotient (TQ)   See *tolerance quotient (TQ)* under *biological indices*.

tolerance range   Range of one or more environmental conditions (i.e., the highest and lowest values) in which an organism can function and survive.

top of the bank   Point on a bank that corresponds to the high water mark for normal streamflows.

topography   Configuration of a surface including its relief and the position of its natural and artificial features.

topset bed   Horizontal sedimentary bed formed at the top of a delta over the existing streambed.

top width   See *top width* under *dimensions*.

torrent   Refers to a violent high streamflow condition that is created by heavy rainfall or rapid snowmelt and is characterized by a near bank-full discharge or greater, increased velocity, standing waves, and high sediment load in alluvial or meltwater streams. See *alluvial stream* and *meltwater stream* under *stream*.

tortuous meander   See *tortuous meander* under *meander*.

tortuous meander channel   See *tortuous meander channel* under *channel pattern*.

total dissolved solids (TDS)   Measure of inorganic and organic materials dissolved in water that pass through a 0.45 μm filter, expressed as mg/L. Sometimes used as an indicator of potential production in habitat quality indices. Also referred to as filterable residue (FR). See *conductivity;* compare to *suspended solids*.

total sediment discharge   Total quantity of sediment passing a stream crosssection during a prescribed unit of time.

total solar radiation   See *total solar radiation* under *solar radiation*.

total storage   Volume of water in a reservoir at any stage from full pool to dead storage.

total stream power   See *total stream power* under *energy*.

total suspended solids (TSS)   Organic and inorganic material left on a standard glass filter of 0.45 μm after a water sample is passed through the filter. Also referred to as filterable residue (FR).

toxicant   (1) A substance that, through its chemical or physical action, kills, injures, or impairs an organism. (2) Any environmental condition that results in a harmful biological effect.

toxicity   Quality, state, or degree of a harmful effect in organisms that results from alteration of natural environmental conditions.

training   See *training* under *remote sensing*.

training wall   See *training wall* under *habitat enhancements*.

tranquil flow   See *tranquil flow* under *flow*.

transect   (1) A line on the ground along which observations are made or data are collected at fixed intervals. (2) A line across a region selected to show spatial relationships of landforms, vegetation, or other features. (3) A straight line across a stream channel, perpendicular to the flow, along which habitat features such as depth or substrate are measured at pre-determined intervals.

transitional habitat   Habitat serving as a boundary between two dissimilar habitat types.

transition flow   See *transition flow* under *flow*.

transition region   Stream reach where the flow changes from laminar to turbulent.

transition zone   Narrow or broad zone where a change occurs from wetlands to nonwetlands.

translation   See *translation* under *wave*.

transmissivity   Rate at which water is transmitted through a unit width of an aquifer under a unit hydraulic gradient. Transmissivity values can be expressed as square meters per day (m^2/d) or square meters per second (m^2/s).

transparency   Ability of water to transmit visible light.

transpiration   Process in plants where water is released as vapor (primarily through the stomata, or pores, in leaves) into the atmosphere.

transportation zone   See *transportation zone* under *fluvial*.

transport capacity   Ability of a stream to transport a suspended sediment load that is expressed as the total weight of sediment.

transport velocity   Velocity required to move sediment of different sizes in a stream.

Transport velocities for various
sizes of streambed materials

Diameter		Transport velocity	
mm	in	cm/s	ft/s
0.005–0.5	0.00002–0.002	15–20	0.49–0.66
0.25–2.5	0.01–0.10	30–65	0.98–2.13
5.0–15	0.2–0.6	80–120	2.62–3.94
25–75	1.0–3.0	140–240	4.59–7.87
100–200	4.0–7.8	270–390	8.86–12.80

transverse bar   See *transverse bar* under *bar*.

transverse flow   See *transverse flow* under *flow*.

transverse rib   Lines of large clasts across a channel that are usually one or two diameters wide.

trash collector   See *trash collector* under *habitat enhancements*.

trash rack   See *trash rack* under *control structure*.

tree retards   See *tree retards* under *habitat enhancements*.

trellis stream   See *trellis stream* under *stream*.

trench   See *trench* under *slow water, pool, scour pool* under the main heading *channel unit*.

tributary   Stream that flows into or joins a larger stream. Synonymous with feeder stream or side stream. Tributary types are based on watershed geomorphology.

lower valley wall tributary   Characterized by moderately steep gradients that occur at the slope break between valley wall and valley floor.

terrace tributary   Stream flowing across terraces to the main stem, originating from springs or from tributaries draining valley slopes.

upper valley wall tributary   Possess very steep gradients, high water velocities, and flow in a stepwise profile of alternating pools and cascades.

wall-based tributary   Flow along the base of a valley wall and into the main stem channel. May flow parallel to the main stem for a short distance.

tripton   Nonliving, nondetritus fragments in water.

trophic   Related to the processes of energy and nutrient transfer (i.e., productivity) from one level of organisms to another in an ecosystem.

autotrophic   Water body where all organic compounds are produced through photosynthesis rather than imported from external sources.

dystrophic   Shallow lake with colored water, high humic and total organic matter content, low nutrient availability, high oxygen demand, and limited bottom fauna. Oxygen is continually depleted and pH is usually low. The stage between a eutrophic lake and a wetland in lake succession.

eutrophic   Water body that is rich in nutrients, organic materials, and productivity. During the growing season, chlorophyll concentrations are typically 10–100 mg/m^3.

heterotrophic   Water body where all organic compounds are derived from sources that are external to the water body.

hypereutrophic   Water body receiving very

high nutrient enrichment, usually from human activities such as agricultural runoff and sewage effluent. Specifically, phosphorus levels are greater than 100 mg/m^3, and chlorophyll levels are greater than 40 mg/m^3.

hypertrophic   See *hypereutrophic*.

hypoeutrophic   Water body with less than the desired nutrients and productivity.

hypotrophic   See *hypoeutrophic*.

mesotrophic   Water body with productivity intermediate between oligotrophic and eutrophic with chlorophyll levels typically at 4–11 mg/m^3 during the growing season.

oligotrophic   Water body characterized by low dissolved nutrients and organic matter, dissolved oxygen near saturation, and chlorophyll levels typically at less than 4 mg/m^3 during the growing season.

status   Position of an organism with respect to energy flow within an aquatic ecosystem.

trophogenic region   Area of a water body where organic production from mineral substances takes place on the basis of light energy and photosynthetic activity.

tropholytic region   Deep area of a water body where decomposition of organic matter predominates.

tropical river   See *tropical river* under **river**.

trough   (1) A long depression or hollow. (2) Small human-built structure of metal, wood, or concrete used to hold water (e.g., watering trough for livestock). See *trench* under *pool* under the main heading **channel unit**.

true color   Color of water resulting from dissolved substances rather than from colloidal or suspended matter.

truncated meander   See *truncated meander* under *meander*.

tumbling flow   See *tumbling flow* under **flow**.

tundra   Zone of low vegetation found above the tree line in the arctic or in mountainous areas but below zones of perpetual ice and snow (i.e., snowbanks or glaciers).

tundra brook   Narrow, shallow stream with low dissolved solids, water temperature usually less than 8° C, and variable flows that are supplied from snow and ice melt.

tundra river   See *tundra river* under **river**.

turbidity   (1) Refers to the relative clarity of a water body. (2) Measure of the extent to which light penetration in water is reduced from suspended materials such as clay, mud, organic matter, color, or plankton. Measured by several nonequivalent standards such as nephelometric turbidity units (NTU), formazin turbidity units (FTU), and Jackson turbidity units (JTU).

turbidity current   A mass of mixed water and sediment that flows downward (sometimes rapidly) along the bottom of an ocean or lake because it is denser than the surrounding water.

turbine   See *turbine* under **control structure**.

turbulence   Streamflows in which the velocity at a given point varies erratically in magnitude and direction and disrupts reaches with laminar flow. Turbulence causes disturbance of the water surface and produces uneven surface levels which results in poor visibility because air bubbles are entrained in the water.

turbulent flow   See *turbulent flow* under **flow**.

turnover   (1) Complete mixing of nutrients and oxygen in a lake that occurs when stratification breaks down due to changes in water temperature, water density, and wind action. (2) Time interval between the use and replacement of one or more nutrients in a nutrient pool. See *fall turnover*, *spring turnover*, *turnover*, and *thermal stratification* under the main heading of **stratification**.

turnover ratio   See *turnover ratio* under **mixing**.

tussock   Clumps or thick tufts of vegetation forming a more solid surface in a wetland.

two-story lake   See *two-story lake* under **lake**.

u

ultrafine particulate matter   See *ultrafine particulate matter* under **organic particles**.

ultranannoplankton   See *ultranannoplankton* under *nannoplankton*.

ultraplankton   See *ultraplankton* under *plankton*.

unchanneled colluvium   See *unchanneled colluvium* under *colluvium* in *valley segments*.

unconfined aquifer   Underground water not held in a confined area as a result of impervious materials that allow movement and interchange of water with adjoining areas.

unconfined channel   See *unconfined channel* under *confinement*.

unconfined meander   See *unconfined meander* under *meander*.

unconsolidated bottom   Bottom in wetlands and deepwater habitats with at least 25% of the area covered by particles smaller than stones and less than 30% covered by vegetation.

unconsolidated bottom wetland   See *unconsolidated bottom wetland* under *wetlands*.

unconsolidated deposits   Sediments with particles that are loosely arranged and are not cemented together that include alluvial, glacial, volcanic, and landslide deposits.

unconsolidated shore wetland   See *unconsolidated shore wetland* under *wetlands*.

undercut bank   See *undercut bank* under *streambank*.

underflow   See *underflow* under *mixing*.

underground water   See *subsurface flow*.

underscour pool   see *underscour pool* under *slow water, pool, scour pool* with the main heading of *channel unit*.

undertow   (1) Any strong current below the surface of a water body that moves in a different direction than the surface current. (2) A seaward subsurface flow or draft of water from waves breaking on an ocean beach.

undifferentiated   See *undifferentiated* under *streambank material*.

uniform flow   See *uniform flow* under *flow*.

unit stream power   See *unit stream power* under *energy*.

universal soil loss equation   An equation that predicts the amount of soil lost in an average year:

$$A = RKLSPC \; ;$$

A = the soil loss in mass per unit area per year;
R = rainfall factor;
K = slope erodibility factor;
L = length of the slope;
S = percent slope;
P = conservation practice factor;
C = cropping and management factor.

unsaturated   Conditions when water can hold additional dissolved gases or other materials or where the subsurface aquifers are able to hold additional water.

unstable areas   Land areas that have a higher probability of increased erosion, landslides, and channel adjustment disturbances during climatic or physical events such as major storms or landslides.

unstable bank   See *unstable bank* under *bank stability* and *streambank*.

unsteady flow   See *unsteady flow* under *flow*.

unsupervised classification   See *unsupervised classification* under *remote sensing*.

upland   (1) Refers to the terrestrial habitat above a water body or any area that is not typically influenced by saturated soil or by standing or moving water. (2) Zone sufficiently above and away from flowing or standing water that is dependent on precipitation for its water supply.

uplift   High land area produced by movements that raise or upthrust underlying rocks in the earth's crust.

upper bank   See *upper bank* under *streambank*.

upper infralittoral   See *upper infralittoral* under *littoral*.

upper valley wall tributary   See *upper valley wall tributary* under *tributary*.

upstream   Direction from which a river or stream flows.

upwelling   Movement of a water mass from the bottom to the surface.

urban runoff   Storm water from city streets that

usually transports litter and organic wastes in addition to water from precipitation.

useable storage   Volume of water in a reservoir that is normally available for domestic water, irrigation, power generation, and other purposes.

UTM (universal transverse mercator)   Grid system for establishing a fixed point between 84° N and 80° S using exact measurements. The planet earth is divided into 60 grids, each 6° longitude by 8° latitude. Line 1 begins at 180° longitude. Line 2 would be at 174°, etc. Latitude lines are designated by letters C through X, but omitting I and O to reduce confusion. Each grid zone is further subdivided into a finer grid of 100,000 square meters. Starting at 180°, the 100,000 m grid blocks are labelled from A to Z, again omitting I and O. Every 18° the lettering starts over. The 100,000 m squares are also lettered from A to V, south to north, beginning at 80° S, again omitting I and O. The sequence repeats every 2,000,000 m, so that the nearest similar grid reference is 101 km or 63 miles. Grid coordinates can be derived to within 10 m.

V

vadose zone   Subsurface zone in a water table where the interstices or spaces in a porous substrate are only partially filled with water.

valley   Elongated, low, and flat area of a landscape, between higher terrain features such as uplands, hills and mountains where runoff and sediment transport occur through downslope convergence.

valley fill   Accumulation of deposits that partially or completely fill a valley.

valley flat   Area in a valley bottom (i.e., floodplain) that becomes inundated under high streamflows. See *floodplain*.

valley floor width index   See *valley floor width index* under *dimensions*.

valley glacier   See *valley glacier* under *glacier*.

valley line   Longitudinal profile of a streambed in a water course. Compare with *thalweg*.

valley morphology   System for classifying valleys based on physical features and profiles.

type I   A V-shaped, confined valley, often structurally controlled and associated with faults. Elevation relief is high and slopes in the bottom are moderately steep or greater than 2%.

type II   Valley with moderate relief and slopes that is relatively stable; gradients less than 4%.

type III   Valley that is primarily depositional, with characteristic debris-colluvial or debris-alluvial fan landforms and slopes in the bottom that are moderately steep or greater than 2%.

type IV   Valley with classic meandering pattern, entrenched or deeply incised, and confined landforms such as canyons or gorges with gentle elevation relief and valley floor gradients less than 2%.

type V   Valley that is glacially scoured with U-shaped trough and valley floor slopes less than 4%.

type VI   Fault-line valley that is structurally controlled and dominated by colluvial slope building processes. Valley floor gradients are often less than 4%

type VII   Valley with a steep to moderately steep landform that is characterized by highly dissected fluvial slopes, typically with deeply incised stream channels, high drainage density, and a very high sediment supply.

type VIII   Valley with multiple river terraces that are positioned laterally along broad valleys with a gentle elevation.

type IX   Glacial outwash plains and dunes formed by deposition in areas with a high sediment supply.

type X   Valley that is very wide with a gentle elevation relief and composed of mostly alluvial materials. Typically this type of valley is found in coastal plains as broad lacustrine or alluvial flats.

type XI   Valley with large river deltas and tidal flats composed of alluvial materials.

valley segments   Valley networks with similar geologic features that are formed through geomorphic processes.

alluvium   Valley that is characterized by fluvial transport of sediment over a predominantly alluvial valley fill.

bedrock   Valley with little soil or sediment and dominated by bedrock.

colluvium   Valley where colluvial fills accumulate and are periodically eroded during rare hydrologic events.

channeled colluvium   Valleys that contain low order streams immediately downstream from unchanneled colluvial valleys.

unchanneled colluvium   Headwater valley segments lacking recognizable stream channels.

estuarine   Valley with a transition zone between terrestrial and marine environments.

valley wall   Portion of a valley slope that is located above a valley flat and relic terraces. In some situations where fans or deltas have been formed, the valley wall is absent.

varves   Sedimentary layers of a lake bed that are deposited at regular intervals, usually one or two times per year. More particularly, the differential sediment deposited in glacial lakes by glacial streams; coarser layers are deposited in summer and finer layers are deposited in winter.

vector   See *vector* under *remote sensing*.

vector data   See *vector data* under *remote sensing*.

vector format   See *vector format* under *remote sensing*.

vegetation   Refers to plant life or total plant cover of an area.

vegetation density   The extent of an area covered by vegetation that is generally expressed as a percentage of a specific area. Also, the percentage of an area at the surface, midwater, and bottom along a vertical transect or at other fixed points in a water body that is occupied by aquatic plants.

vegetation layer   Subunit of a plant community where all component species exhibit the same growth form (e.g., trees, saplings, or herbs).

vegetation–soil rating   Used to evaluate the impacts of consumptive and nonconsumptive animal activity on riparian vegetation and impacts to soils from erosion.

vegetative cover   In aquatic systems, vegetation that provides cover for protection of fish and other aquatic organisms (e.g., algal mats, macrophytes, and overhanging riparian vegetation).

velocity   Speed at which water travels downstream. More specifically, the time rate of motion calculated as the distance traveled divided by the time required to travel that distance and expressed as cm/s, m/s, or ft/s.

critical velocity   (1) Maximum water velocity in which a fish can sustain its position for a specified length of time. For example, the maximum forward swimming speed that a fish can sustain over a specified distance or length of time. (2) Velocity in a channel when flow changes from laminar to turbulent.

fish velocity or focal point velocity   Velocity at the location occupied by a fish that is measured at the fish's snout. Synonymous with snout velocity or facing velocity.

mean column velocity   Average velocity of water measured on an imaginary vertical line at any point in a stream. A measurement at 60% of the maximum depth for depths less than 76 cm (30 in), measured from the surface, closely approximates the average velocity for the water column. In water depths greater than 76 cm (30 in), the average of measurements made at 20% and 80% of the depth approximates the mean column velocity.

mean cross section velocity   Mean velocity of water flowing in a channel at a given cross section that is equal to the discharge divided by the cross section of the area.

swimming velocity   See *swimming speed*.

thalweg velocity   Mean velocity of a water column in a stream that is measured along the thalweg.

vernal lake   See *vernal lake* under *lake*.

vernal pond   See *vernal pond* under *pond*.

vertical gully side   See *vertical gully side* under *gully side form*.

vertical stability   Indication of the net effect of deposition or scour of a streambed in a specific reach over a long period of time that is described as degrading or aggrading.

vertical velocity profile   A parabolic line that describes water velocity in a vertical plane of a stream channel at a given location. The velocity along the line varies from zero at the bottom to a maximum value near the surface.

very long duration   Duration of inundation for a single flood event that is greater than one month.

viscosity   Adhesive quality of a liquid including water that exhibits resistance to friction.

viscous water   Water to which a thickening agent has been added to reduce surface runoff.

v-notch   (1) Narrow ravine or valley with steep sides and a V-shaped cross section that usually contains a watercourse. (2) Type of weir with a V-shaped notch used for gaging discharge in small streams.

volcanic   See *volcanic* under **streambank material**.

volcanic lake   See *volcanic lake* under **lake**.

volume   Mass or quantity of water enclosed within a specific water body that is reported as cubic meters (m³), cubic feet (ft³), or acre-feet (ac-ft). See also *volume* under **large organic debris**.

volume curve   Graph with depth on the vertical axis and percentage volume along the horizontal axis.

volume development   See *volume development* under **dimensions**.

W

wadi   Channel of a water course that is dry except during periods of rainfall.

wake   Track of waves from objects such as a boat moving through the water.

wall-based pond   See *wall-based pond* under **pond**.

wall-based tributary   See *wall-based tributary* under **tributary**.

wandering meander channel   See *wandering meander channel* under **channel pattern**.

warm monomictic   See *warm monomictic* under **mixing**.

warm polymictic   See *warm polymictic* under **mixing**.

warmwater fishes   A broad term applied to fish species that inhabit waters with relatively warm water temperatures (optimum temperatures generally between 15–27°C (60–80°F)). Compare with *coldwater fishes, coolwater fishes*.

warmwater lake   See *warmwater lake* under *lake*.

wash   (1) Dry bed of an elongated depression or channel that is formed by water in an intermittent stream. Often associated with an arid environment that is characterized by flash flooding, high bed load movement, and sparse vegetation. (2) A general pattern of erosion, movement, or exposure to moving water.

washing   Removal of fines from surface substrate materials by wave action or flowing water.

washload   See *washload* under **sediment load**. Compare with *suspended load* under **sediment load**.

wastewater   Water carrying dissolved or suspended solids generated by human activities.

water   Colorless, odorless, tasteless, slightly compressible liquid, composed of two atoms of hydrogen and one of oxygen, that is a major constituent of all living matter. Occurs as several forms including rain, sleet, snow, and ice that cycles through atmospheric, surface, and subsurface transport. Generic word used in place of lake, reservoir, river, or other water body. See *water body*.

water balance   A record of outflow from, inflow to, and storage in a hydrologic unit such as an aquifer or drainage basin. See **aquifer, drainage basin**.

water bar   Shallow ditch excavated across a road at an angle to collect and divert surface water from a roadway to prevent erosion of the road bed. Also referred to as cross ditch.

water body   Any natural or artificial pond, lake, stream, river, estuary, or ocean that contains permanent, semi-permanent, or intermittent standing or flowing water.

water budget   The balance of all water moving into and out of a specified area within a specified period of time.

water column   Portion of water in a water body extending vertically from a given point on the surface to any depth; generally used to locate, describe, or characterize the chemical and physical constituents at a given depth or depth range.

water course   Natural or artificial channel with perennial or intermittent water and definable bed and banks.

water cycle   Large-scale circulation of water between the atmosphere and earth that involves the processes of precipitation, condensation, runoff, evaporation, and transportation. See *hydrologic cycle*.

waterfall   See *falls* under *fast water—turbulent* under the main heading **channel unit**.

water level fluctuation   Annual vertical fluctuation of the surface in a water body including reservoirs and lakes.

water logging   (1) Generally refers to the saturation of soil with water which causes inadequate aeration may be detrimental to plants. (2) Process or condition where the water table reaches or rises above the ground surface or its capillary fringe is near to the surface.

water mark   Visible line on upright structures along a water body of debris accumulation, sediment deposits, or stain that represents the maximum height of water.

water pressure   Pressure in water created by gravity resulting from elevational differences. See *head*.

water quality   Term used to describe biological, chemical, and physical characteristics of an aquatic environment, usually in relation to the uses of water.

water regime   Refers to the presence and pattern of surface water at a given location.

   intermittently exposed   Surface water is present throughout the year except in years of extreme drought.

   intermittently flooded   Substrate is usually exposed but covered with surface water for variable periods of time.

   permanently flooded   Water covers the land surface throughout the year in all years and where vegetation includes obligate hydrophytes.

saturated   Substrate is saturated to the ground surface for extended periods of time during the growing season but surface water is seldom present.

seasonally flooded   Surface water is present for extended periods of time, especially early in the growing season, but is absent by the end of the season for most years. Generally occurs in areas where the water table is near the surface.

semipermanently flooded   Surface waters persist throughout the growing season in most years where the water table is at or near the surface in all years.

temporarily flooded   Surface water is present for brief periods during the growing season but the water table is usually well below the ground surface.

water right   Authorization by prior ownership, contract, purchase, or appropriation, to use water for designated purposes.

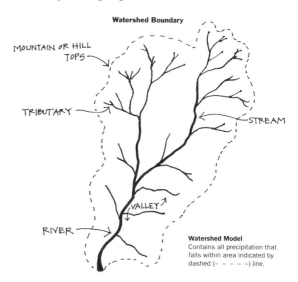

watershed (from Firehock and Doherty 1995)

watershed   (1) Region or area drained by surface and groundwater flow in rivers, streams, or other surface channels. A smaller watershed can be wholly contained within a larger watershed. (2) The divide between two catchment areas. See *drainage area*.

water spreading   Diversion of streamflow, generally from a watercourse onto a gently sloping and porous ground to conserve water, maintain or increase plant growth, reduce flood peaks, and replenish or recharge groundwater.

water-stable aggregate   Soil aggregate that is stable to the action of water.

water storage   The retention of water in a specified location by natural or artificial means.

water table   Depth below which the ground is saturated with water; generally expressed as linear depth below the soil surface to the upper layer of groundwater.

waterway   River, canal, or other navigable channel that is used as a route or way of travel or transport.

water year   See *climatic year*.

water yield   Total outflow from all or a part of a drainage basin through surface runoff or subsurface aquifers within a given time (i.e., a year). Also referred to as annual water yield.

wave   A vertical disturbance on the surface of a stream, river, lake, or sea in the form of a ridge or swell that is usually caused by wind.

  antinode   Where the maximum vertical movement occurs during a rocking motion in a water body.

  breaker   Collapse of waves in front of asymmetrical wave.

  capillary waves   Refers to water that is flowing in small waves. See *ripple* under *wave*.

  gravity waves   Short surface waves with a wavelength greater than 6.28 cm.

  height   Perpendicular distance between a wave crest and trough.

  internal progressive wave   Horizontal water movements associated with shearing flow at the metalimnion–epilimnion interface.

  internal seiche   Layers in a lake of differing density that oscillate relative to one another. See *seiche* under *wave*.

  Kelvin wave   Rotary wave in a lake that is always parallel to a shoreline.

  langmuir circulation   Elongate spirals of water that rotate about a horizontal axis parallel to the water surface and wind direction. See *streaks*.

  length   Distance between two successive wave crests.

  long wave   Wave with a length that is long in comparison to, and much greater than, the water depth. A long wave is nondispersive and travels at a speed that is independent of its wavelength.

  nodal point   Point where no vertical movement occurs during a rocking motion in water.

  oscillation   Fluctuating wave in deep water where each particle of water vacillates for a short distance as the wave moves forward and creates a wave motion that is very different from regular wave motion.

  period   The elapsed time between the arrival of two successive wave crests at any point.

  period of uninodal surface oscillation (t)   Defined as:

  $$t = 2L/(gz)^{\frac{1}{2}};$$

    L = the length of the basin at the surface;
    z = mean depth of the basin;
    g = acceleration of gravity (980.6 cm/s).

  plunging breaker   The forward fall of a wave becomes convex and the crest curls over but collapses with insufficient depth to complete a vortex.

  Poincare wave   Wave occurring only in very large lakes where a cellular system of gyres or circular currents with alternating upward and downward oscillations of wave centers results in the wave direction rotating clockwise once every wave cycle.

  progressive wave   See *surface wave* under *wave.*

  refraction   Process where a series of waves in shallow water that are moving at an angle to the shoreline change direction to become parallel with the bottom contours.

  ripple   Waves less than 6.28 cm long that move with a slight rise and fall or ruffling of the water surface. See *capillary waves* under *wave.*

  seiche   Rhythmic motions of a lake, bay, or other water body resulting in fluctuations in water level that are generally caused by wind but may be caused by geologic processes such as earthquakes or tectonic movements of plates. A seiche is formed by a long, standing wave that oscillates from the surface to some depth of a lake or landlocked sea that varies in period from a few minutes to several hours.

significant wave height   Average height from trough to crest of one-third of the largest waves.

spilling breaker   The forward collapse of a wave that spills downward over the front of another wave.

standing wave   Permanent, nonmoving water created by deflection of swiftly flowing water by large woody material, large boulders, or the bank. Also applied to the wave associated with a hydraulic jump.

surface seiche   Long standing wave that attains maximum amplitude at the surface and shoreline and results in oscillating water levels along the shore of a water body.

surface wave   Vertical movement of water from wind action on the surface of a water body.

swell   Long, massive, and crestless wave (or succession of waves) that is generally produced by wind and often continues after its cause. The distance between swells is much longer than the distance between waves in a general wave pattern.

translation   Horizontal (i.e., forward) movement of water that is covered with surface waves and that is characteristic of shallow water.

wave velocity   Speed of wave travel that is calculated by dividing the wave length by the period of time between waves.

wave breaker   Structure that is designed and constructed of solid materials to intercept and dissipate the energy of waves in areas such as the entrance to a harbor. See *breakwall*.

wave cut platform   Gently sloping surface produced by wave erosion that may extend some distance into a lake or sea from the base of a cliff.

wave velocity   See *wave velocity* under *wave*.

weathering   In situ decomposition and disintegration of bedrock from both chemical and physical processes.

wedge dam   See *wedge dam* under *habitat enhancements*.

weed   Troublesome or noxious plant, often exotic or introduced, that grows profusely in the wild and often is well-adapted to quickly invade and thrive in disturbed areas.

weighted usable area (WUA)   (1) An index of the capacity of a stream reach to support a particular species and life stage; expressed as the actual area or the percentage of suitable habitat area that is available per unit length of a stream at a given flow. (2) Total surface area having a certain combination of hydraulic and substrate conditions multiplied by the composite suitability index of use by fish or other aquatic species for the specific combination of conditions at a given flow.

weir   (1) Notch or depression in a levee, dam, embankment, or other barrier across or bordering a stream that regulates or measures the flow of water. (2) Barrier constructed across a stream to guide fish into a trap. See also *weir* under *habitat enhancements*.

well   Hole dug or drilled into the earth to obtain water, or a natural spring source of water that is in a deep depression.

well head   Outlet, or fountainhead, to move water to the surface of the earth.

wetland boundary   Point on the land surface where a transition or shift occurs from wetlands to nonwetlands or aquatic habitat.

wetland hydrology   Total of wetness characteristics in an area that is inundated by water or has saturated soils for sufficient duration to support hydrophytic vegetation. Such permanent or periodic inundation by water or prolonged soil saturation generally results in anaerobic soil conditions.

wetlands   Land areas that are wet at least for part of the year, are poorly drained, and are characterized by hydrophytic vegetation, hydric soils, and wetland hydrology. See *wetland hydrology*.

adjacent   Wetlands separated from other aquatic habitats by constructed dikes or barriers, natural river berms, beach dunes, or similar features.

aquamarsh   Water body where the original open water area is nearly or completely obscured by emergent and floating aquatic vegetation.

aquatic bed   Class of wetland and deepwater habitat dominated by plants that grow principally on or below the surface of the water for most of the growing season in most years.

artificial wetland   Wetlands created, intentionally or accidentally, by the anthropogenic activities of humans.

backswamp   An extensive, marshy, depressed area in floodplains between the natural levee borders of channels and valley sides or terraces.

blanket bog   Large upland areas, typically in cold, wet climates, where extensive accumulations of undecomposed peat cover waterlogged, nutrient-poor ground. Also referred to as blanket mire.

bog   (1) Wet, spongy land that is usually poorly drained, highly acidic, and nutrient-rich, and is characterized by an accumulation of poorly to moderately decomposed peat and surface vegetation of mosses and shrubs. (2) A peatland dominated by ericacenous shrubs, sedges, and moss with a saturated water regime, or a forested peatland dominated by evergreen trees and larch.

carr   Wetland on organic soil with greater than 25% cover of shrubs, typically dominated by willow.

cienagas   Wetland associated with spring and seep systems in isolated arid basins of the Southwest.

emergent wetland   Class of wetland habitat characterized by erect, rooted, herbaceous hydrophytes, excluding mosses and lichens, that are present for most of the growing season.

fen   Low-lying peatland that is partially covered with relatively fast moving, nutrient-rich, neutral to basic water that is rich in calcium. Fens are often dominated by sedges and rushes that form peat when they die and decay.

forested wetland   Wetland habitat characterized by woody vegetation (conifer and deciduous species) that is 6 m (20 ft) tall or taller, and that occurs near springs, seeps, or areas with natural high water tables.

freshwater marsh   Area with continuously water-logged soil that is dominated by emergent herbaceous plants but without a surface accumulation of peat.

fringe marsh   Saturated, poorly drained area, intermittently or permanently water covered, that is close to or along the edge of a land mass.

haline marsh   Saturated, poorly drained seashore area that is intermittently or permanently water covered and covered with aquatic and grass-like vegetation.

high moor   Type of bog where vegetation and peat are low in nutrients, minerals, and nitrogen and that is located in sites with a cool, humid climate (as in higher latitudes) where heavy precipitation has leached most of the nutrients from the soil and caused waterlogging for much of the year, creating a blanket bog or blanket peat.

lotic (riparian) wetland   Lotic wetlands are associated with flowing water found along streams, rivers, and drainages. They have a defined channel, an associated floodplain, continuously transport water, and include seeps, springs, beaver ponds, and wet meadows.

low moor   Type of bog or swamp composed of peat or mulch soil, formed in eutrophic or mesotrophic waters that are relatively rich in minerals (generally the drainage from a surrounding catchment area into a basin that may have been a former lake).

mangrove swamp   Swampy or tidal area dominated by tropical trees or shrubs of the genus *Rhizophora* with exposed interlacing adventitious roots above ground.

marsh   Water-saturated, poorly drained wetland area that is periodically or permanently inundated to a depth of 2 m (6.6 ft) and that supports an extensive cover of emergent, nonwoody vegetation, without peat-like accumulations.

moor   An open, uncultivated tract of land with a peat-like soil that supports low vegetation, typically of coarse grass and sedges, in low elevations and with sphagnum and cotton "grass" at higher and wetter elevations. At its wettest, a moor is similar to a bog.

morass   Tract of low, soft, wet ground.

moss–lichen wetland   Wetland dominated by mosses (mainly peat moss) and lichens with little tall vegetation.

muskeg   A bog composed of deep accumulations of organic material in wet, poorly drained boreal areas (often with permafrost), that is usually dominated by *sphagnum* mosses, often contains tussocks, and has shrubby plants, and small, scattered trees.

nonpersistent wetland   Wetland dominated by plants that fall on the surface or below the water surface at the end of the growing season when there is no obvious sign of emergent vegetation.

palustrine   Nontidal wetland that is dominated by trees, shrubs, persistent emergents, mosses, or lichens.

peat bog   Bog with a dominant underlying material of peat. See *bog* under *wetlands*.

pocosin   Local term for swamp or bog in the coastal plain of the southeastern United States.

problem area wetland   Wetland that is difficult to identify because it may lack indicators of wetland hydrology and hydric soils; usually dominated by plant species that are characteristic of nonwetlands.

quaking bog   Dense, interwoven accumulation of organic matter and living organisms that floats on open water in the littoral zone of dystrophic lakes.

raised bog   A bog containing an accumulation of organic matter, but low in plant nutrients, with excellent capillarity that raises the water level in the mat. As a result, the central portion of the bog extends above the natural groundwater level.

reference wetland   Wetland within a relatively homogeneous biogeographic region that is representative of a specific hydrogeomorphic wetland type.

riverine wetland   Any wetland or deepwater habitat contained within a stream channel.

saline marsh   Saturated, poorly drained inland area, intermittently or permanently water covered, containing various dissolved salts, and with aquatic and grass-like vegetation. See *saline* under *salinity*.

salt marsh   Low areas adjacent to saline springs or lakes covered by salt-tolerant vegetation or similar areas near a sea that are periodically flooded by seawater but not exposed to daily tides. See *tidal marsh* under *wetlands*.

scrub–shrub wetland   Wetland that includes areas dominated by low, woody vegetation less than 6 m (20 ft) tall that are stunted because of existing environmental conditions.

soligenous fen   Peatland formed in waters draining, at least partially, areas of high mineral content.

swamp   Tree- or tall shrub-dominated wetlands that are characterized by periodic flooding and nearly permanent subsurface water flow through mixtures of mineral sediments and organic materials without peat-like accumulation.

tidal marsh   Low, flat marshland often traversed by interlaced channels and tidal sloughs that are subject to oceanic tides; vegetation consists of salt-tolerant bushes and grasses. See *salt marsh* under *wetlands*.

unconsolidated bottom wetland   Wetlands that have bottoms with at least 25% cover of particles smaller than stones and a vegetative cover of less than 30%.

unconsolidated shore wetland   Wetlands that have unconsolidated substrates with less than 75% of the area covered by rocky material, less than 30% of the area covered by vegetation other than pioneering plants, and water regimes that include irregular flooding or saturation.

wet meadow   Meadows characterized by wet soils that are normally waterlogged within a few inches of the ground surface and slow surface and subsurface flows. Channels are typically poorly defined or nonexistent and vegetation is dominated by grasses and riparian-dependent species.

wooded swamp   Wetland dominated by trees (i.e., a forested wetland).

wetland soil   Soil that is saturated for prolonged periods of time and that is accompanied by anaerobic conditions. Hydric soils that are sufficiently wet to support hydrophytic vegetation are also considered to be wetland soils.

wetland status   Refers to wetland species that have exhibited an ability to develop to maturity and reproduce in an environment where all or portions of the soils within the root zone become, periodically or continuously, saturated with water during the growing season.

facultative wetland species   Species usually occurring in wetlands but occasionally found in nonwetlands.

obligate wetland species   Species that almost always occur under the natural conditions of wetlands.

wetland vegetation   Vegetation that occurs in areas with hydric soils and wetland hydrology. See *wetland, wetland hydrology*.

wet line   Length of sounding line between a water surface and the bottom of a water body.

wet meadow   See *wet meadow* under *wetlands*.

wetted cross section   See *wetted cross section* under *dimensions*.

wetted perimeter   (1) Distance along the bottom and sides of a cross section in a body of water in contact with water that is roughly equal to the width plus two times the mean depth. (2) See also *wetted perimeter* under *dimensions*.

wetted width   See *wetted width* under *dimensions*.

wet water   Water to which a surfactant has been added to reduce the surface tension.

white water   Occurs where flows are sufficiently fast and turbulent to entrain numerous air bubbles in the water, giving the water a whitish cast.

width   See *width* under *dimensions*.

width : depth ratio   See *width : depth ratio* under *dimensions*.

windfall   Trees or parts of trees felled by high winds. See *blowdown* under *large organic debris*.

windrow   (1) A row or line of material swept together by the wind. (2) An accumulation of road fill or surfacing material left on the road shoulders from poor grading. (3) Woody materials from timber sales or land-clearing activities that are shoved into elongated piles.

windthrow   See *blowdown* under *large organic debris*.

wind turbidity   Turbidity resulting from agitation of bottom sediments from wind action, usually in shallow areas where waves mobilize fine sediments.

windward shore   The shore toward which the wind blows.

wing dam   See *check dam* under *habitat enhancements*.

wing wall   (1) Part of a dam that is constructed into the bank. (2) Extension of a bridge abutment that is constructed to retain fill material in the road bed to prevent it from entering the watercourse.

winter heat income   See *winter heat income* under *heat budget*.

winterkill   The death of fishes or other organisms in a water body during a prolonged period of ice and snow cover. Death is usually caused by oxygen depletion due to the lack of photosynthesis.

Wisconsin bank cover   See *Wisconsin bank cover* under *habitat enhancements*.

wooded swamp   See *wooded swamp* under *wetlands*.

woody debris   See *woody debris* under *large organic debris*.

WUA   Acronym for weighted usable area. An index of physical habitat for a specific species.

▶ X

xeric   Locations lacking in water due to limited rainfall.

xerophytic   Plant species that are typically adapted for life in conditions where a lack of water is a limiting factor for growth and reproduction. These species are capable of growth in extremely dry conditions as a result of morphological, physiological, and reproductive adaptations.

▶ Y

yield   To produce or give forth a product or the quantity of a product by a natural process or cultivation. The harvest, actual or estimated, of living organisms, expressed as numbers, weight, or as a proportion of the standing crop, for a given period of time. Also refers to the amount of water produced from a given drainage or watershed. See *drainage, watershed*.

young river See *young river* under *river*.

▶ **Z**

zone (1) Area characterized by similar flora or fauna. (2) Belt or area to which certain species are limited.

zone of aeration Soil and capillaries above the water table that are exposed to air from the atmosphere. See *capillary* and *soil*.

zone of influence Area contiguous to a ditch, channel, or other drainage feature that is directly affected by it.

zone of saturation The soil zone that is located below the permanent water table.

zooplankton See *zooplankton* under *plankton*.

References

Aird, P. L. 1994. Conservation for the sustainable development of forests worldwide: a compendium of concepts and terms. Forestry Chronicle 70:666–674.

Allan, J. D. 1995. Stream ecology: structure and function of running waters. Chapman and Hall, London.

Allee, W. C., A. E. Emerson, O. Park, T. Park, and K. P. Schmidt. 1949. Principles of animal ecology. Saunders, Philadelphia. (Reprinted 1967.)

Alonso, C. V., and S. T. Combs. 1990. Stream bank erosion due to bed degradation—a model concept. Transactions of the American Society for Applied Engineering 33:1239–1248.

Amoros, C. 1991. Changes in side-arm connectivity and implications for river system management. Rivers 2:105–112.

Angermeier, P. L., and J. R. Karr. 1986. Applying an index of biotic integrity based on stream-fish communities: considerations in sampling and interpretation. North American Journal of Fisheries Management 6:418–429.

Armantrout, N. B., editor. 1982. Acquisition and utilization of aquatic habitat inventory information. American Fisheries Society, Western Division, Bethesda, Maryland.

Armantrout, N. B. 1983. A flexible integrated aquatic habitat inventory and monitoring system. Pages 428–431 in J. F. Beel and T. Atterbury, editors. Renewable resource inventories for monitoring changes and trends: an international conference. Oregon State University, College of Forestry, Corvallis.

Armour, C. L., K. P. Burnham, and W. S. Platts. 1983. Field methods and statistical analyses for monitoring small salmonid streams. U.S. Fish and Wildlife Service, Office of Biological Services, FWS/OBS/83/33, Washington, D.C.

Arnette, J. L. 1976. Nomenclature for instream assessments. Pages 9–15 in C. B. Stalnaker and J. L. Arnette, editors. Methodologies for the determination of stream resource flow requirements: an assessment. U.S. Fish and Wildlife Service, Office of Biological Services, Washington, D.C.

Bair, E. S. 1995. Hydrogeology. Pages 285–310 in A. D. Ward and J. Elliot, editors. Environmental hydrology. Lewis Publishers, Boca Raton, Florida.

Balon, E. K. 1982. About the courtship rituals in fishes, but also about a false sense of security given by classification schemes, 'comprehensive' reviews and committee decisions. Environmental Biology of Fishes 7:193–197.

Bancroft, B., and K. Kelsey. 1994. "In-stream" management techniques to enhance fish habitat. Pages 61–69 in N. B. Federicton, editor. Proceedings of a symposium on riparian management. Canadian Forest Service, Maritime Region, Nova Scotia.

Barraclough, C. L. No date. Glossary of terms used in aquatic inventory. Prepared by Aqua-Tex Scientific Consulting, Ltd., for British Columbia Department of the Environment, Resource Inventory Committee, Victoria.

Bates, R. L., and J. A. Jackson. 1980. Glossary of geology, 2nd edition. American Geological Institute, Washington, D.C.

Bayha, K. 1981. Glossary of terms pertinent to instream flow work. (Mimeographed draft manuscript.) Cooperative Instream Flow Service Group, Western Energy and Land Use Team, U.S. Fish and Wildlife Service, Fort Collins, Colorado.

Bayley, P. B. 1991. The flood pulse advantage and restoration of river- floodplain systems. Regulated Rivers Research & Management 6:75–86.

Bell, M. C. 1986. Fisheries handbook of engineering requirements and biological criteria. U.S. Army Corps of Engineers, North Pacific Division, Fish Passage Development and Evaluation Program, Portland, Oregon.

Bell, M. C. 1990. Fisheries handbook of engineering requirements and biological criteria. U.S. Army Corps of Engineers, North Pacific Division, Fish Passage Development and Evaluation Program, Portland, Oregon.

Benda, L., T. J. Beechie, R. C. Wissmar, and A. Johnson. 1992. Morphology and evolution of salmonid habitats in a recently deglaciated river basin, Washington State, USA. Canadian Journal of Fisheries and Aquatic Sciences 49:1246–1256.

Beschta, R. L. 1978. Inventorying small streams and channels on wildland watersheds. Pages 104–113 *in* Proceedings of a national workshop on integrated inventories of renewable natural resources. U.S. Forest Service General Technical Report RM-55.

Beschta, R. L., and W. L. Jackson. 1979. The intrusion of fine sediments into a stable gravel bed. Journal of the Fisheries Research Board of Canada 36:204–210.

Beschta, R. L., S. J. O'Leary, R. E. Edwards, and K. D. Knoop. 1981. Sediment and organic matter transport in Oregon Coast Range streams. Water Resources Institute Publication WRRI-70, Corvallis, Oregon.

Beschta, R. L., and W. S. Platts. 1986. Morphological features of small streams: significance and function. Water Resources Bulletin 22:369–379.

Binns, N. A., and F. M. Eiserman. 1979. Quantification of fluvial trout habitat in Wyoming. Transactions of the American Fisheries Society 108:215–228.

Bisson, P. A., and D. Montgomery. 1996. Valley segments, stream reaches, and channel units. Pages 23–52 *in* R. F. Hauer and G. A. Lamberti, editors. Methods in stream ecology. Academic Press, New York.

Bisson, P. A., J. L. Nielsen, R. A. Palmason, and L. E. Grove. 1982. A system of naming habitat types in small streams, with examples of habitat utilization by salmonids during low streamflow. Pages 62–73 *in* N. B. Armantrout, editor. Acquisition and utilization of aquatic habitat inventory information. American Fisheries Society, Western Division, Bethesda, Maryland.

Boehne, P. L., and R. A. House. 1983. Stream ordering: a tool for land managers to classify western Oregon streams. U.S. Department of the Interior, Bureau of Land Management, Technical Note T/N:OR-3, Portland, Oregon.

Bovee, K. D. 1978. The incremental method of assessing habitat potential for coolwater species, with management implications. Pages 340–346 *in* R. L. Kendall, editor. Selected coolwater fishes of North America. American Fisheries Society, Special Publication 11, Bethesda, Maryland.

Bovee, K. D., and T. Cochnauer. 1977. Development and evaluation of weighted criteria, probability-of-use curves for instream flow assessments: fisheries. U.S. Fish and Wildlife Service, Federal Interagency Energy/Environment Research and Development Program, Office of Research and Development, U.S. Environmental Protection Agency, FWS/OBS-77-63, Fort Collins, Colorado.

Brussock, P. P., A. V. Brown, and J. C. Dixon. 1985. Channel form and stream ecosystem modes. Water Resources Bulletin 21:859–866.

Bryant, M. D., P. E. Porter, and S. J. Paustian. 1990. Evaluation of a stream channel-type system for southeast Alaska. U.S. Forest Service General Technical Report PNW-267.

Bryant, M. D., B. E. Wright, and B. J. Davies. 1992. Application of a hierarchical habitat unit classification system: stream habitat and salmonid distribution in Ward Creek, southeast Alaska. U.S. Forest Service Research Note PNW-508.

Chamberlin, T. W. 1980. Aquatic system inventory (biophysical stream surveys). British Columbia Ministry of Environment, Assessment and Planning Division, Technical Paper 1, Victoria.

Chamberlin, T. W. 1980. Aquatic survey terminology. British Columbia Ministry of Environment, Assessment and Planning Division, Technical Paper 2, Victoria.

Cherry, J., and R. L. Beschta. 1989. Coarse woody debris and channel morphology: a flume study. Water Resources Bulletin 25:1031–1036.

Cheslak, E. F., and A. S. Jacobson. 1990. Integrating the instream flow incremental methodology with a population response model. Rivers 1:264–288.

Chutter, F. M. 1972. An empirical biotic index of the quality of water in South African streams and rivers. Water Research 6:19–30.

Cole, G. A. 1983. Textbook of limnology, 3rd edition. C. V. Mosby, St. Louis, Missouri.

Collotzi, A. W. 1977. A systematic approach to the stratification of the valley bottom and the relationship to land use planning. Internal paper. U.S. Forest Service, Bridge-Teton National Forest, Jackson Hole, Wyoming.

Collotzi, A. W. 1986. GAWS: general aquatic wildlife system. Proceedings of the annual meeting of the Colorado-Wyoming Chapter, American Fisheries Society 21:113–114. (Fort Collins, Colorado.)

Collotzi, A. W., and D. K. Dunham. 1978. Inventory and display of aquatic habitats. Pages 533–542 in A. Marmelstein, editor. Classification, inventory, and analysis of fish and wildlife habitat. U.S. Fish and Wildlife Service, Office of Biological Services, FWS/OBS-78/76, Washington, D.C.

Committee on Restoration of Aquatic Ecosystems. 1992. Restoration of aquatic ecosystems. Committee on Restoration of Aquatic Ecosystems, National Research Council. National Academy Press, Washington, D.C.

Cooperrider, A. Y., R. J. Boyd, and H. R. Stuart. 1986. Inventory and monitoring of wildlife habitat. U.S. Department of the Interior, Bureau of Land Management, Denver.

Cowardin, L. M., V. Carter, F. Golet, and E. J. Laroe. 1979. Classification of wetlands and deepwater habitats of the United States. U.S. Fish and Wildlife Service, Office of Biological Services, FWS/OBS-79-31.

Cronquist, A. 1961. Introductory Botany. Harper and Row, New York. 892 pp.

Crouch, R. J. 1987. The relationship of gully sidewall shape to sediment production. Australian Journal of Soil Research 25:531–539.

Cuplin, P., C. Armour, R. Corning, D. Duff, and A. Oakley. 1974. Fisheries and aquatic habitat inventory and analysis techniques. U.S. Department of the Interior, Bureau of Land Management, Technical Note, Denver, Colorado.

Deason, W. O. 1975. Environmental glossary. U.S. Department of the Interior, Bureau of Reclamation, Washington, D.C.

DeLeeuw, A. D. 1982. Guide to aquatic survey terminology and data entry procedures and codes. Canadian Ministry of Environment, Aquatic Studies Branch, Vancouver.

Densmore, A. L., R. S. Anderson, B. G. McAdoo, and M. A. Ellis. 1997. Hillslope evolution by bedrock landslides. Science 275:369–372.

Dodge, D. P., G. A. Goodchild, I. MacRitchie, J. C. Tilt, and D. G. Waldriff. 1984. Manual of instructions: aquatic habitat inventory surveys. Canadian Ministry of Natural Resources, Fisheries Branch, Official Procedural Manual Policy Fl.2.03.01., Ottawa.

Dolloff, C. A., D. G. Hankin, and G. H. Reeves. 1993. Basin wide estimation of habitat and fish populations in streams. U.S. Forest Service, Southeast Forest Experiment Station General Technical Report SE-83, Asheville, North Carolina.

Duff, D. D. 1982. General aquatic wildlife system (GAWS). In R. Wiley, D. Bartschi, A. Binns, D. Duff, J. Erickson, and B. Platts, editors. Proceedings of the Rocky Mountain stream habitat management workshop. American Fisheries Society, Western Division, Jackson Hole, Wyoming.

Dunham, D. K., and A. Collotzi. 1975. The transect method of stream habitat inventory—Guidelines and applications. U.S. Forest Service, Ogden, Utah.

Dunster, J., and K. Dunster. 1996. Dictionary of natural resource management. University of British Columbia Press, Vancouver.

Government Institutes, Inc. 1991. Natural resources glossary. Government Institutes, Inc., Rockville, Maryland.

Elliot, W. J., and A. D. Ward. 1995. Soil erosion and control practices. Pages 117–204 in A. D. Ward and J. Elliot, editors. Environmental hydrology. Lewis Publishers, Boca Raton, Florida.

Enviro Control, Inc. 1980. Reach file phase II report: a standardized method for classifying status and type of fisheries. Draft Report to U.S. Fish and Wildlife Service, Western Energy and Land Use Team, Fort Collins, Colorado.

ERDAS (Earth Resources Data Analysis System). 1993. ERDAS field guide, 3rd edition. ERDAS, Inc., Atlanta.

Estes, C. C., and D. S. Vincent-Lang. 1984. Aquatic habitat and instream flow investigations. Draft Report 3. Alaska Power Authority, Anchorage, Alaska.

Fajen, O. F., and R. E. Wehnes. 1982. Missouri's method of evaluating stream habitat. Pages 117–123 in N. B. Armantrout, editor. Acquisition and utilization of aquatic habitat inventory information. American Fisheries Society, Western Division, Bethesda, Maryland.

Fausch, K. D., C. L. Hawkes, and M. G. Parsons. 1988. Models that predict standing crop of stream fish habitat variables: 1950–85. U.S. Forest Service General Technical Report PNW-GTR-213.

Fausch, K. D., J. Lyons, J. R. Karr, and P. L. Angermeier. 1990. Fish communities as indicators of environmental degradation. Pages 123–144 *in* S. M. Adams, editor. Biological indicators of stress in fish. American Fisheries Society, Symposium 8, Bethesda, Maryland.

Federal Interagency Committee for Wetland Delineation. 1989. Federal manual for identifying and delineating jurisdictional wetlands. U.S. Army Corps of Engineers, U.S. Environmental Protection Agency, U.S. Fish and Wildlife Service, and U.S. Department of Agriculture, Soil Conservation Service, Washington, D.C.

Ferguson, R. I., K. L. Prestegaard, and P. J. Ashworth. 1989. Influence of sand on hydraulics and gravel transport in a braided gravel bed river. Water Resources Research 25:635–643.

Ferren, W. R., Jr., P. L. Fiedler, R. A. Leidy, K. D. Lafferty, and L. A. K. Mertes. 1996. Wetlands of California: part II: classification and description of wetlands of the central and southern coastal watersheds. Madrono 43:125–182.

Firehock, K., and J. Doherty. 1995. A citizen's streambank restoration handbook. Save Our Streams, Izaak Walton League of America, Inc., Gaithersburg, Maryland.

Ford-Robertson, F. C., editor. 1971. Terminology of forest science technology practice and products. The Multinational Forestry Terminology Series 1 (English language version). Society of American Foresters, Bethesda, Maryland.

Forest Ecosystem Management Assessment Team. 1993. Forest ecosystem management: an ecological, economic, and social assessment. U.S. Forest Service, Portland, Oregon.

Foth, H. D. 1984. Fundamentals of soil science. Wiley, New York.

Fredriksen, R. L., and R. D. Harr. 1979. Soil, vegetation, and watershed management of the Douglas-fir region. Pages 231–260 *in* P. E. Heilman, H. Anderson, and D. M. Baumgartner, editors. Forest soils of the Douglas-fir region. Washington State University, Cooperative Extension Service, Pullman.

Frick, G. W. 1980. Environmental glossary. Government Institutes, Inc., Washington, D.C.

Frissell, C. A., and W. J. Liss. 1986. Classification of stream habitat and watershed systems in south coastal Oregon. Progress Report. Oak Creek Laboratory of Biology, Department of Fisheries and Wildlife, Oregon State University, Corvallis.

Gas, I. G., P. J. Smith, and R. C. L. Wilson, editors. 1972. Understanding the Earth, 2nd edition. The Massachusetts Institute of Technology Press, Cambridge.

Gomez, B., and M. Church. 1989. An assessment of bed load sediment transport formulae for gravel bed rivers. Water Resources Research 25:1161–1186.

Gordon, N. D., T. A. McMahon, and B. L. Finlayson. 1992. Stream hydrology. Wiley, New York.

Gosse, J. C., and W. T. Helm. 1982. A method for measuring microhabitat components for lotic fishes and its application with regard to brown trout. Pages 138–149 *in* N. B. Armantrout, editor. Acquisition and utilization of aquatic habitat inventory information. American Fisheries Society, Western Division, Bethesda, Maryland.

Grant, G. E., and F. J. Swanson. 1995. Morphology and processes of valley floors in mountain streams in western Cascades Oregon. Pages 83–101 *in* J. E. Costa, A. J. Miller, K. W. Potter, and P. R. Wilcock, editors. Natural and anthropogenic influences in fluvial geomorphology. American Geophysical Union, Geophysical Monograph 89, Washington, D.C.

Grant, G. E., F. J. Swanson, and M. G. Wolman. 1990. Pattern and origin of stepped-bed morphology in high-gradient streams, Western Cascades, Oregon. Geological Society of America Bulletin 102:340–352.

Gray, D. H., and A. T. Leiser. 1982. Biotechnical slope protection and erosion control. Van Nostrand Reinhold, New York.

Greentree, W. J., and R. C. Aldrich. 1976. Evaluating stream trout habitat on large-scale aerial color photographs. U.S. Forest Service Research Paper PSW-123/1976.

Ham, D. 1996. Aerial photography and videography standards: application for stream inventory and assessment. Resource Inventory Committee, Fisheries Branch, Ministry of Environment, Lands and Parks, Victoria.

Hamilton, K., and E. P. Bergersen. 1984. Methods to estimate aquatic habitat variables. Prepared by the Colorado Cooperative Fishery Research Unit, Fort Collins, for the U.S. Department of the Interior, Bureau of Reclamation, Denver, Colorado.

Hankin, D. G., and G. H. Reeves. 1988. Estimating total fish abundance and total habitat area in small streams based on visual estimation methods. Canadian Journal of Fisheries and Aquatic Sciences 45:834–844.

Hansen, E. C., G. R. Alexander, and W. H. Dunn. 1983. Sand sediment in a Michigan trout stream. Part I. A technique for removing sand bedload from a stream. North American Journal of Fisheries Management 3:355–364.

Hansen, P. L., and five coauthors. 1995. Classification and management of Montana's riparian and wetland sites. Montana Forest and Conservation Experiment Station, Miscellaneous Publication 54.

Hasfurther, V. R. 1985. The use of meander parameters in restoring hydrologic balance to reclaimed stream beds. Pages 21–40 *in* J. A. Gore, editor. The restoration of rivers and streams: theories and experience. Butterworth, Stoneham, Massachusetts.

Hawkins, C. P., and 10 coauthors. 1993. A hierarchical approach to classifying stream habitat features. Fisheries 18(6):3–12.

Heede, B. H. 1976. Gully development and control: the status of our knowledge. U.S. Forest Service Research Paper RM-169.

Heede, B. H. 1980. Stream dynamics: an overview for land managers. U.S. Forest Service, General Technical Report RM-72.

Helm, W. T., J. C. Gosse, and J. Bich. 1982. Life history, microhabitat and habitat evaluation systems. Pages 150–153 *in* N. B. Armantrout, editor. Acquisition and utilization of aquatic habitat inventory information. American Fisheries Society, Western Division, Bethesda, Maryland.

Helm, W. T. 1985. Aquatic habitat inventory: glossary and standard methods. American Fisheries Society, Western Division, Bethesda, Maryland.

Herrington, R. B., and D. K. Dunham. 1967. A technique for sampling general fish habitat characteristics of streams. U.S. Forest Service Research Paper INT-41.

Hilsenhoff, W. L. 1977. Use of arthropods to evaluate water quality of streams. Wisconsin Department of Natural Resources Technical Bulletin 100, Madison.

Hilton, S., and T. E. Lisle. 1993. Measuring the fraction of pool volume filled with fine sediment. U.S. Forest Service Research Note PSW-RN-414.

Hogle, J. S., T. A. Wesche, and W. A. Hubert. 1993. A test of the precision of the habitat quality index model II. North American Journal of Fisheries Management 13:640–643.

Hubert, W. A., and S. J. Kozel. 1993. Quantitative relations of physical habitat features to channel slope and discharge in unaltered mountain streams. Journal of Freshwater Ecology 8:177–183.

Hunt, C. E., and V. Huser. 1988. Down by the river: the impacts of federal water projects and policies on biological diversity. Island Press, Covelo, California.

Hynes, H. B. N. 1970. The ecology of running waters. University of Toronto Press, Toronto.

Hynes, H. B. N. 1983. Groundwater and stream ecology. Hydrobiologia 100:93–99.

Independent Scientific Group. 1996. Return to the river: restoration of salmonid fishes in the Columbia River ecosystem. Development of an alternative conceptual foundation and review and synthesis of science underlying the Columbia River Basin Fish and Wildlife Program of the Northwest Power Planning Council, Portland, Oregon. Prepublication Draft. NWPPC 96-6.

Johnson, S. W., and J. Heifetz. 1985. Methods for assessing effects of timber harvest on small streams. NOAA (National Oceanic and Atmospheric Administration) Technical Memorandum NMFS (National Marine Fisheries Service) F/NWC-73.

Joweth, I. 1993. A method for objectively identifying pool, run, and riffle habitats for physical measurements. New Zealand Journal of Marine and Freshwater Research 27:241–248.

Judson, S., K. S. Deffeyes, and R. B. Hargraves. 1976. Physical geology. Prentice-Hall, Englewood Cliffs, New Jersey.

Junk, W. J., P. B. Bayley, and R. E. Sparks. 1989. The flood pulse concept in river floodplain systems. Canadian Special Publication of Fisheries and Aquatic Sciences 106:110–127.

Kappesser, G. B. 1993. Riffle stability index. Draft Report. U.S. Forest Service, Idaho Panhandle National Forest, Coeur d'Alene, Idaho.

Karr, J. R. 1981. Assessment of biotic integrity using fish communities. Fisheries 6(6):21–27.

Karr, J. R., P. R. Yant, and K. D. Fausch. 1987. Spatial and temporal variability of the index of biotic integrity in three midwestern streams. Transactions of the American Fisheries Society 116:1–11.

Kauffman, J. B., R. L. Beschta, N. Otting, and D. Lytjen. 1997. An ecological perspective of riparian and stream restoration in the western United States. Fisheries 22(5):12–24.

Kaufmann, P., and E. G. Robison. 1993. A quantitative habitat assessment protocol for field evaluation of physical habitat in small wadable streams. Draft version. Department of Forest Engineering and Department of Fisheries and Wildlife, Oregon State University, and U.S. Environmental Protection Agency, Environmental Research Laboratory, Corvallis.

Kellerhalls, R., M. Church, and D. I. Bray. 1976. Classification and analysis of river processes. Journal of the Hydraulics Division, American Society of Civil Engineers, 102:813–829.

Kent, C., and J. Wong. 1982. An index of littoral zone complexity and its measurement. Canadian Journal of Fisheries and Aquatic Sciences 39:847–853.

Kershner, J. L., W. M. Snider, D. M. Turner, and P. B. Moyle. 1992. Distribution and sequencing of mesohabitats: are there differences at the reach scale? Rivers 3:179–190.

Kinsolving, A. D., and M. B. Bain. 1990. A new approach for measuring cover in fish habitat studies. Journal of Freshwater Ecology 5:373–378.

Kohler, C. C., and W. A. Hubert, editors. 1993. Inland fisheries management in North America. American Fisheries Society, Bethesda, Maryland.

Kondolf, G. M., G. F. Cada, and M. J. Sale. 1987. Assessing flushing-flow requirements for brown trout spawning gravels in steep streams. Water Resources Bulletin 23:927–935.

Kondolf, G. M., and E. R. Micheli. 1995. Evaluating stream restoration projects. Environmental Management 19:1–15.

Lanka, R. P., W. A. Hubert, and T. A. Wesche. 1987. Relations of geomorphology to stream habitat and trout standing stock in small Rocky Mountain streams. Transactions of the American Fisheries Society 116:21–28.

Leonard, P. M., and D. J. Orth. 1986. Application and testing of an index of biotic integrity in small, coolwater streams. Transactions of the American Fisheries Society 115:401–414.

Leopold, L. B., and T. Maddock, Jr. 1953. The hydraulic geometry of stream channels and some physiographic implications. U.S. Geological Survey, Professional Paper 252, Washington, D.C.

Leopold, L. B., M. G. Wolman, and J. P. Miller. 1964. Fluvial processes in geomorphology. Freeman, San Francisco.

Lienkaemper, G. W., and F. J. Swanson. 1987. Dynamics of large woody debris in streams in old-growth Douglas-fir forests. Canadian Journal of Forest Research 17:150–156.

Lisle, T. 1987. Using "residual depth" to monitor pool depths independently of discharge. U.S. Forest Service, Pacific Southwest Forest and Range Experiment Station, Research Paper PSW-394, Berkeley, California.

Lisle, T. 1987. Overview: channel morphology and sediment transport in steepland streams. Proceedings of the symposium on erosion and sediment transport in the Pacific Rim. International Association of Hydrologic Sciences 165:287–297.

Lisle, T. E. 1995. Particle size variation between bed load and bed material in natural gravel bed channels. Water Resources Research 31:1107–1118.

Lisle, T. E., and S. Hilton. 1992. The volume of fine sediment in pools: an index of sediment supply in gravel-bed streams. Water Resources Bulletin 28:371–383.

Lotspeich, F. B., and F. H. Everest. 1981. A new method for reporting and interpreting textural composition of spawning gravel. U.S. Forest Service Research Note PNW-369.

Lyons, J. 1992. Using the index of biotic integrity (IBI) to measure environmental quality in warmwater streams of Wisconsin. U.S. Forest Service General Technical Report NC-149.

Maciolek, J. A. 1978. Insular aquatic ecosystems: Hawaii. Pages 103–120 in A. Marmelstein, editor. Classification, inventory, and analysis of fish and wildlife habitat. U.S. Fish and Wildlife Service, Office of Biological Services, FWS/OBS-78/76, Washington, D.C.

Marmelstein, A. 1978. Classification, inventory, and analysis of fish and wildlife habitat. U.S. Fish and Wildlife Service, Office of Biological Services, FWS/OBS-78/76, Washington, D.C.

Matthews, J. E. 1969. Glossary of aquatic ecological terms. U.S. Department of the Interior, Federal Water Pollution Control Administration, Ada, Oklahoma.

McCain, M., L. Decker, and K. Overton. 1990. Stream habitat classification and inventory procedures for northern California. U.S. Forest Service, Pacific Southwest, Region 5, FHR #1, San Francisco.

Meehan, W. R., editor. 1991. Influences of forest and rangeland management on salmonid fishes and their habitats. American Fisheries Society Special Publication 19, Bethesda, Maryland.

Mish, F. C., editor 1985. Webster's ninth new collegiate dictionary. Merriam-Webster, Inc., Springfield, Massachusetts.

Montgomery, D. R., and J. M. Buffington. 1993. Channel classification, prediction of channel response, and assessment of channel condition. Washington Department of Natural Resources, Report TFW(Timber-Fish-Wildlife)-SH10-93-002, Olympia.

Morisawa, M. 1968. Streams: their dynamics and morphology. McGraw-Hill, New York.

Murphy, M. L., and five coauthors. 1987. The relationship between stream classification, fish and habitat in southeast Alaska. U.S. Forest Service, Alaska Region, Wildlife and Habitat Management Note 12, Juneau.

Myers, T. J., and S. Swanson. 1991. Aquatic habitat condition, stream type, and livestock bank damage in northern Nevada. Water Resources Bulletin 27:667–677.

Nawa, R. K., and C. A. Frissell. 1993. Measuring scour and fill of gravel streambeds with scour chains and sliding-bead monitors. North American Journal of Fisheries Management 13:634–639.

Nelson, R. W., G. C. Horack, and J. E. Olson. 1978. Western reservoir and stream habitat improvement handbook. U.S. Fish and Wildlife Service, Office of Biological Services, Western Land Use and Energy Team, Fort Collins, Colorado.

Oberdorff, T., and R. M. Hughes. No date. Modification of an index of biotic integrity based on fish assemblages to characterize rivers of the Seine Basin, France. NSI (Northrop Services, Inc.) Technology Services Corporation, Corvallis, Oregon.

Odum, E. P., and H. T. Odum. 1959. Fundamentals of ecology. Saunders, Philadelphia.

Olson, J. E., E. A. Whippo, and G. C. Hovak. 1981. Reach file phase II report: a standardized method for classifying status and type of fisheries. U.S. Fish and Wildlife Service, Office of Biological Services, FWS/OBS-81/31, Washington, D.C.

Orth, D. J. 1983. Aquatic habitat measurements. Pages 61–84 in L. A. Nielsen and D. L. Johnson, editors. Fisheries techniques. American Fisheries Society, Bethesda, Maryland.

Osborn, J. F. 1982. Estimating spawning habitat using watershed and channel characteristics (a physical systems approach). Pages 154–160 in N. B. Armantrout, editor. Acquisition and utilization of aquatic habitat inventory information. American Fisheries Society, Western Division, Bethesda, Maryland.

Osborne, L. L., and six coauthors. 1992. Influence of stream location in a drainage network on the index of biotic integrity. Transactions of the American Fisheries Society 121:635–643.

Overton, C. K., J. D. McIntyre, R. Armstrong, S. L. Whitwell, and K. A. Duncan. 1995. User's guide to fish habitat: descriptions that represent natural conditions in the Salmon River Basin, Idaho. U.S. Forest Service General Technical Report INT-GTR-322.

Parsons, S., and S. Hudson. 1985. Channel cross section surveys and data analysis. U.S. Department of the Interior, Bureau of Land Management, Report TR-4341-1, Denver, Colorado.

Paustian, S. J., D. Perkison, D. A. Marich, and P. Hunsicker. 1983. An aquatic value rating procedure for fisheries and water resource management in southeast Alaska. Pages 17-1 to 17-29 in Managing Water Resources for Alaska's Development. Water Resources Association, Alaska Section, Fairbanks.

Payne, N. F., and F. Copes, technical editors. 1986. Wildlife and fisheries habitat improvement handbook. U.S. Forest Service, Wildlife and Fisheries Administrative Report (unnumbered), Washington, D.C.

Pennak, R. W. 1978. The dilemma of stream classification. Pages 59–66 in A. Marmelstein, editor. Classification, inventory, and analysis of fish and wildlife habitat. U.S. Fish and Wildlife Service, Office of Biological Services, FWS/OBS-78/76, Washington, D.C.

Pfankuch, D. J. 1975. Stream reach inventory and channel stability evaluation. U.S. Forest Service, Northern Region, Missoula, Montana.

Platts, W. S. 1974. Methodology for classifying aquatic environments in mountainous lands for entry into land use planning. Presented at the William F. Sigler Symposium, Utah State University, Logan.

Platts, W. S., W. F. Megahan, and G. W. Minshall. 1983. Methods for evaluating stream, riparian, and biotic conditions. U.S. Forest Service, Intermountain Forest and Range Experiment Station, Ogden, Utah.

Press, F., and R. Siever. 1978. Earth. Freeman, San Francisco.

Rabe, F. W., C. Eizinga, and R. Breckenridge. 1994. Classification of meandering glide and spring stream natural areas in Idaho. Natural Areas Journal 14:188–202.

Rahel, F. J., and W. A. Hubert. 1991. Fish assemblages and habitat gradients in Rocky Mountain—Great Plains streams: biotic zonation and additive patterns of community change. Transactions of the American Fisheries Society 120: 319–332.

Rassam, G. N., J. Gravesteijn, and R. Potenza. 1988. Multilingual thesaurus of geosciences. Pergamon Press, New York.

Rechard, P. A., and R. McQuisten. 1968. Glossary of selected hydrologic terms. Wyoming Water Resources Research Institute, Water Resources Series 1 (Revised), University of Wyoming, Laramie.

Renard, K. G., J. M. Laflen, G. R. Foster, and D. K. McCool. 1994. The revised universal soil loss equation. Pages 105–124 in R. Lal, editor. Soil erosion research methods. Soil and Water Conservation Society, St. Lucie Press, Ankeny, Iowa.

Rinne, J. N. 1985. Physical habitat evaluation of small stream fishes: point vs transect, observation vs capture methodologies. Journal of Freshwater Ecology 3:121–131.

Robison, E. G., and R. L. Beschta. 1990. Coarse woody debris and channel morphology interactions for undisturbed streams in southeast Alaska, USA. Earth Surface Processes and Landforms 15:149–156.

Roseboom, D. 1994. Case studies on biotechnical stream bank protection. Proceedings of the annual meeting of the Forestry Committee, Great Plains Agriculture Council, Manhattan, Kansas. Great Plains Agriculture Council Publication 149:57–65.

Rosgen, D. L. 1994. A classification of natural rivers. Catena 22:169–199.

Rosgen, D. 1996. Applied river morphology. Wildland Hydrology, Pagosa Springs, Colorado.

Rosgen, D., and B. L. Fittante. 1986. Fish habitat structures—a selection guide using stream classification. Pages 163–179 in J. G. Miller, J. A. Arway, and R. F. Carline, editors. Proceedings, 5th trout stream habitat improvement workshop. Pennsylvania Fish Commission, Harrisburg.

Ruttner, F. 1953. Fundamentals of limnology. University of Toronto Press, Toronto.

Schumm, S. A. 1993. River response to baselevel change: implications for sequence stratigraphy. Journal of Geology 101:279–294.

Sedell, J. R., J. E. Richey, and F. J. Swanson. 1989. The river continuum concept: a basis for the expected ecosystem behavior of very large rivers. Canadian Special Publication of Fisheries and Aquatic Sciences 106:49–55.

Shera, W. P., and D. J. Grant. 1980. A hierarchical watershed coding system for British Columbia. Canadian Ministry of Environment, Research Analysis Branch of Technology, Paper 3, Vancouver.

Sherrets, H. D. 1989. Wildlife watering and escape ramps on livestock water developments: suggestions and recommendations. U.S. Department of the Interior, Bureau of Land Management, Technical Bulletin 89-4, Boise, Idaho.

Shirazi, M. A., and W. K. Seim. 1979. A stream system evaluation—an emphasis on spawning habitat for salmonids. U.S. Environmental Protection Agency Report EPA-800/3-79-109, Corvallis, Oregon.

Shirazi, M. A., W. K. Seim, and D. H. Lewis. 1981. Characterization of spawning gravel and stream system evaluation. Pages 227–278 in Proceedings, conference on salmon spawning gravel: a renewable resource in the Pacific Northwest? Washington State University, Water Research Center Report 39, Pullman.

Simonson, T. D., J. Lyons, and P. D. Kanehl. 1994. Quantifying fish habitat in streams: transect spacing, sample size, and a proposed framework. North American Journal of Fisheries Management 14:607–615.

Stauffer, J. C., and R. M. Goldstern. 1997. Comparison of three qualitative habitat indices and their applicability to prairie streams. North American Journal of Fisheries Management 17:348–361.

Stone, E., III. 1996. B.L.M. balloon photography. U.S. Department of the Interior, Bureau of Land Management, Technical Paper, Oregon State Office, Portland.

Storer, T. I., and R. L. Usinger. 1957. General zoology. McGraw-Hill, New York.

Strahler, A. N. 1957. Quantitative analysis of watershed geomorphology. Transactions of the American Geophysical Union 38:913–920.

Stream Enhancement Research Committee. 1980. Stream enhancement guide. Canadian Ministry of Environment, Department of Fisheries and Oceans, Vancouver.

Stream Systems Technology Center. 1993. Would the real bankful stand up! Stream Notes (April 1993), Stream Systems Technology Center, Fort Collins, Colorado.

Sullivan, K., T. E. Lisle, C. A. Dolloff, G. E. Grant, and L. M. Reid. 1987. Stream channels: the link between forests and fishes. Pages 39–97 in E. O. Salo and T. W. Cundy, editors. Streamside management: forestry and fishery interaction. University of Washington, Institute of Forest Resources, Contribution 57, Seattle.

Swanson, F. J., and six coauthors. 1987. Mass failures and other processes of sediment production in Pacific Northwest forest landscapes. Pages 9–38 in E. O. Salo and T. W. Cunday, editors. Streamside management: forestry and fishery interaction. University of Washington, Institute of Forest Resources, Contribution 57, Seattle.

Terrene Institute. 1994. Riparian road guide: managing roads to enhance riparian areas. Terrene Institute, Washington, D.C.

Thorp, J. H., and A. P. Covich. 1991. Ecology and classification of North American freshwater invertebrates. Academic Press, San Diego, California.

U.S. Army Corps of Engineers. 1987. Corps of Engineers wetland delineation manual. U.S. Army Corps of Engineers, Environmental Laboratory Wetlands Research Program, Technical Report Y-87-1, Vicksburg, Mississippi.

U.S. Fish and Wildlife Service. 1980. Ecological services manual—habitat evaluation procedures (HEP). U.S. Fish and Wildlife Service, Division of Ecological Services, Ecological Services Manual ESM102, Washington, D.C.

U.S. Fish and Wildlife Service. 1981. Ecological services manual—standards for the development of habitat suitability index models. U.S. Fish and Wildlife Service, Division of Ecological Services, Ecological Services Manual ESM103, Washington, D.C.

U.S. Forest Service. 1985. Fisheries habitat evaluation handbook (monitoring). U.S. Forest Service, Region 6, FSH 2609.23, Portland, Oregon.

U.S. Forest Service. 1990. Fish habitat assessment handbook. Draft report. U.S. Forest Service, Region 5, San Francisco.

U.S. Geological Survey. 1977 with updates. National handbook of recommended methods for water-data acquisition. U.S. Geological Survey, Reston, Virginia.

Valett, H. M., S. G. Fisher, and E. H. Stanley. 1990. Physical and chemical characteristics of the hyporheic zone of a Sonoran desert stream. Journal of the North American Benthological Society 9:201–215.

Van Pelt, J., M. J. Woldenberg, and R. W. H. Verwer. 1989. Two generalized topological models of stream network growth. Journal of Geology 97:281–299.

Vannote, R. L., G. W. Minshall, K. W. Cummins, J. R. Sedell, and C. E. Cushing. 1980. The river continuum concept. Canadian Journal of Fisheries and Aquatic Sciences 37:130–137.

Wakeley, J. S., and L. J. O'Neil. 1988. Techniques to increase efficiency and reduce effort in applications of the habitat evaluation procedures (HEP). U.S. Army Corps of Engineers, Environmental Impact Research Program, Technical Report EL-86-33, Washington, D.C.

Ward, J. V., and J. A. Stanford. 1995. Ecological connectivity in alluvial river ecosystems and its disruption by flow regulation. Regulated Rivers 11:105–119.

Washington Forest Practice Board. 1992. Standard methodology for conducting watershed analysis. Washington Forest Practice Board Manual Version 1,10. Washington Forest Practice Board, Washington State Department of Natural Resources, Olympia.

Webster, J. R., S. W. Golladay, E. F. Banfield, D. J. D'Angelo, and G. T. Peters. 1990. Effects of forest disturbance on particulate organic matter budgets of small streams. Journal of the North American Benthological Society 9: 120–140.

Welcomme, R. L. 1985. River fisheries. FAO (Food and Agriculture Organization of the United Nations) Fisheries Technical Paper 262.

Welcomme, R. L. 1989. Floodplain fisheries management. Pages 209–233 in J. A. Gore and G. E. Petts, editors. Alternatives in regulated river management. CRC Press, Boca Raton, Florida.

Wesche, T. A. 1973. Parametric determination of minimum stream flow for trout. University of Wyoming, Water Resources Research Institute, Water Resources Series 37, Laramie.

Wesche, T. A. 1985. Stream channel modifications and reclamation structures to enhance fish habitat. Pages 103–163 in J. A. Gore, editor. The restoration of rivers and streams: theories and experience. Butterworth, Stoneham, Massachusetts.

Wesche, T. A. 1993. Watershed management and land-use practices. Pages 181–203 in C. C. Kohler and W. A. Hubert, editors. Inland fisheries management in North America. American Fisheries Society, Bethesda, Maryland.

Wetzel, R. G. 1975. Limnology. Saunders, Philadelphia.

White, R. J., and O. M. Brynildson. 1967. Guidelines for management of trout stream habitat in Wisconsin. Wisconsin Department of Natural Resources Technical Bulletin 39.

Williams, O. R., R. B. Thomas, and R. L. Daddow. 1988. Methods for collection and analysis of fluvial sediment data. U.S. Forest Service Report WSTG-TP-00012, Washington, D.C.

Winget, R. N., and F. A. Mangum. 1979. Biotic condition index: integrated biological, physical and chemical stream parameters for management. U.S. Forest Service, Intermountain Region, Ogden, Utah.

Winters, R. K., editor. 1977. Terminology of forest science technology practice and products. Addendum One. Society of American Foresters, Bethesda, Maryland.

Wood-Smith, R. D., and J. M. Buffington. 1996. Multivariate geomorphic analysis of forest streams: implications for assessment of land use impacts on channel condition. Earth Surface Processes and Landforms 21:377–393.

Yeou-Koung T., and W. E. Hathhorn. 1989. Determination of the critical locations in a stochastic stream environment. Ecological Modelling 45:43–61.

Young, M. K., W. A. Hubert, and T. A. Wesche. 1991. Selection of measures of substrate composition to estimate survival to emergence of salmonids and to detect changes in stream substrate. North American Journal of Fisheries Management 11:339–346.